T0230502

The Science of Human Evolution

John H. Langdon

The Science of Human Evolution

Getting it Right

John H. Langdon
University of Indianapolis
Indianapolis, IN, USA

ISBN 978-3-319-82389-8 ISBN 978-3-319-41585-7 (eBook)
DOI 10.1007/978-3-319-41585-7

Softcover reprint of the hardcover 1st edition 2016

Printed on acid-free paper

This Springer imprint is published by Springer Nature
The registered company is Springer International Publishing AG Switzerland

Acknowledgments

I wish to thank my students, my friends and colleagues Richard Smith and Zach Throckmorton and also Mikaela Bielawski, and anonymous reviewers for the helpful feedback. And, as always, I am grateful for the constant support of Terry Langdon in all I do.

Contents

Introduction: The Method of Science

Abstract The scientific method is the best tool our society possesses to generate knowledge and understanding of the natural world. In practice, it is sometimes hindered by human prejudice and error and the difficulty of abandoning one idea for a better one. The case studies in this book examine how science has been practiced in the field of paleoanthropology, how scholars were misled into errors, and how, eventually, they got it right.

Every schoolchild is taught the basic steps of the scientific method: observe, hypothesize, test through experiments, and then reevaluate the hypothesis. Real practice is much more complex. These steps may occur in any order or simultaneously. "Experiments" often are not conducted in a laboratory setting and take many forms. Hypotheses may be interwoven with intuition, implicit assumptions, and errors, although repeated testing of hypotheses is expected to weed these out. Constructing and testing hypotheses is often difficult, but proving hypotheses correct is usually impossible. Scientists must always be aware of the possibility that more complete explanations may come along.

The scientific method has proved to be a powerful tool for acquiring knowledge. It has been adopted throughout the social and historical sciences and applied for such disparate purposes as authenticating authorship of manuscripts, solving crimes, and investigating new teaching strategies. Many of these fields may suffer at times from "physics envy," the desire for straightforward natural laws that define clear cause-and-effect relationships. On close examination, the natural world is not so tidy and unambiguous. In both physics and chemistry, more so in the life sciences, and especially as we study behavioral sciences, laws turn into probabilities. We can predict how populations or particles or organisms respond on average or how individuals are likely to behave under certain circumstances, but particular events occur in the context of myriad variables that are difficult to know. Certainty becomes nearly impossible when we attempt to study human beings.

Anthropologists have struggled throughout the discipline to identify universals of human behavior and society. Among its many branches, physical anthropology makes the strongest claim to be a natural science. By viewing humans as animals and primates, it attempts to apply the same methods for studying our anatomy, physiology, ecology, genetic constitution, and evolutionary origins as biologists would for any organism. This is not achieved without a struggle and may have many false starts and blind ends.

It is the purpose of this book to illuminate this struggle and, in doing so, to shed light on the nature, process, and limitations of natural science. The case studies in this volume span the history of paleoanthropology, from the early nineteenth century to the present. They show successes as well as failures so that we may learn from both.

The Scientific Method

Science begins with observations. It is based on empirical observations of the physical universe, which constitute data. Data are gathered with the senses—if not by naked eyes or ears, then through some instrumentation or secondary effect. Electron microscopes, DNA sequencers, unmanned spacecraft, and measures of isotopes are extensions of our senses. If science is grounded in observation, then its subject matter must be limited to physical objects and events. People collect observations throughout their lives—in everyday experience and in being taught what others have observed. Each person will filter, sort, and evaluate these observations and use them to construct a personal understanding of how the universe operates, the individual's worldview. Eventually, however, science depends on disciplined observation that is systematic and objective. For example, our second study considers William Pengelly, who created a method of excavation that preserves critical observations of context for fossils and artifacts pulled from the ground. His name is little known, but his system of grids and recording is now the starting point for modern field archaeology.

Textbooks tell us that observations lead to hypothesis. Ultimately that is true, but many hypotheses come from other hypotheses. In the classic but apocryphal story, Isaac Newton thought of gravity when an apple fell on his head. Such "Aha!" moments occurring out of context are rare, but it is true that Isaac Newton had frequently observed objects falling and incorporated those observations into his worldview. What set him apart from everyone else is that he asked "Why?" and then attempted an answer.

Charles Darwin's revelation occurred over decades. He began with conventional religious beliefs about creation plus some unconventional but poorly formed ideas about evolution that were circulating among naturalists in the early 1800s. His famous voyage around the world on the *H.M.S. Beagle* opened his eyes to many aspects of natural history that the current model could not explain. With continued thought and study, he merged ideas from biology, geology, and economics and created a new paradigm that ushered in what we call the Darwinian Revolution (Case Study 1).

More commonly, hypotheses are inspired by the work of others. For example, Ernst Haeckel, who is introduced in the third case study, was inspired by Darwin and incorporated Darwin's ideas about evolution into his own worldview to make hypotheses about human origins. Sometimes ideas and technology are borrowed from other disciplines. Many of the major advances within paleoanthropology have come about this way. The field now incorporates knowledge and technologies from archaeology, geology, physics, genetics, ecology, ethnography, animal behavior, and forensic sciences.

Many hypotheses are inductive arguments: empirical observations enable scientists to detect patterns and formulate rules. However, inductive arguments and theories are tentative, because any inductive argument may be threatened by a contrary observation. Likewise, hypotheses may need to be adjusted in the future, to accommodate new data and exceptions. Thus hypotheses and theories are likewise regarded as tentative.

If inductive reasoning is never certain, does it have any value? Induction shows that natural laws, such as Newton's law of gravity, will always hold. This is a principle known as uniformitarianism. If one cannot assume this to be valid, science has no foundation. In our everyday lives, if we cannot make inductions and prediction, there is no basis for our actions. In practice, however, we do act on and build upon such arguments through our technology and through further development of theories. Successful application of inductive hypotheses increases our confidence in them but should not override the need of science to remain open to refinement and improvement of our understanding.

There are two rules by which scientists play that place limits on the natural sciences. First, science can only work with naturalistic explanations. The laws by which the observable universe operates must be explained by invariant properties of that universe. Second, supernatural phenomena lie outside the bounds of science. By definition, supernatural phenomena that do not obey natural laws cannot be objectively observed. Therefore such phenomena cannot be measured, studied, or given a place in the physical universe.

What if scientists were allowed to relax these rules? What if inductive logic is not valid? What if the laws of the universe were different in the past? What if we open the door to supernatural explanations? If these rules are discarded, then science can no longer make predictions. We cannot be certain whether what we observe today has any relation to what we will observe tomorrow. We cannot use empirical knowledge to reconstruct the past or design technology for the future. If we resort to supernatural explanations, we have no way to validate those explanations because they are now divorced from our senses. In short, the scientific method and scientific knowledge become useless.

Does this mean there are no supernatural phenomena? Must we assume there is no God or ghosts or fate? No. Such phenomena are beyond the reach of scientific inquiry or explanation. Literature, ethics, history, and art are also beyond scientific investigation—that is why they are defined as different disciplines of study. These pursuits have different rules and different objectives. They reveal truths and insights of their own, and individuals would be poorer without them. They are different ways of knowing that deserve to sit alongside natural science but not in place of it.

Hypotheses need to be tested if they are to advance from mere speculation to science. We apply the hypothesis to make a prediction ("if this is true then I should observe…."), then set up appropriate conditions, and see whether the predicted observations hold. If so, the hypothesis is affirmed but not proven. If the observations do not match predictions, we need to modify or discard the hypothesis or identify factors that explain the anomaly.

To summarize, science is a method of induction based on observation by which people seek to understand the physical universe. Science can only study physical phenomena and can only invoke naturalistic explanations. Observations and inductive reasoning may be used to generate hypotheses, from which people make predictions about future observations. As they test these, our experiments strengthen or contradict our hypotheses. If these rules are ignored, science is robbed of its value. It is the purpose of this book to examine how science operates in a specific discipline.

The Context of Science

Our observations are interpreted in a theoretical framework of how we understand the world. Our minds must be prepared for what we observe or it will not mean anything to us. For example, what may have been one of the first dinosaur bones known to modern science was sent to Robert Plot, first Keeper of the Ashmolean Museum at Oxford, who published an illustration of it in 1677 (Fig. 1). Although familiar with skeletons of living animals, Plot had no reference to interpret it, and he ascribed it to the thighbone of a giant human. Today, the original specimen has been lost, but from a published illustration, we believe it was the distal femur of a dinosaur called *Megalosaurus*. Although it is easy to laugh at Plot's mistake, he was interpreting the fossil in the context of his understanding of the world, which was influenced by the Biblical passage commonly translated "There were giants in the

NATURAL HISTORY
OF
OXFORD-SHIRE,
Being an Essay toward the Natural History
OF
ENGLAND.
By R. P. LL. D.

Fig. 1 The first dinosaur fossil reported in scientific literature: the distal femur of *Megalosaurus*. Originally published in *Robert Plot (1677) Natural History of Oxfordshire*, Public domain

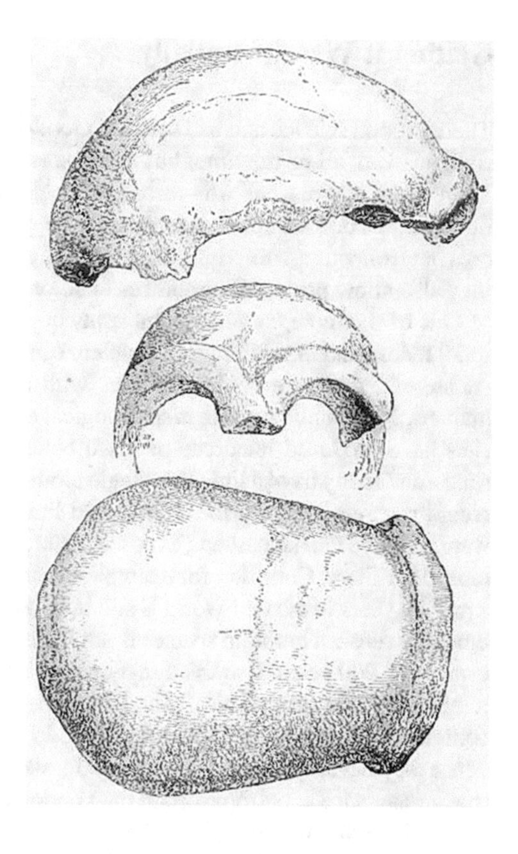

Fig. 2 The original Neanderthal cranium. Source: John G. Rothermel (1894). Fossil Man. Popular Science Monthly 44:616 ff

earth in those days" (Genesis 6:4). We know about dinosaurs from later discoveries, and that knowledge informs our interpretation of Plot's illustration. Thus, it is not sufficient simply to have observations and data; we must also have a context in which they can make sense.

Because we recognize that how we understand our observations may be colored by our worldviews, it is necessary that our observations be accurately recorded and repeatable by other researchers. Inaccurate data is worse than useless because it can be misleading, but when it is possible for other researchers to replicate an experiment, errors can be corrected. The first adult Neanderthal cranium discovered in Gibraltar in 1848 was shelved in the British Museum and forgotten for a century because its discoverers did not have a way to understand it. The second, from Feldhofer Cave in Germany, was understood as a pathological idiot or a member of a primitive human race (Fig. 2). However, in both cases, the fossils were preserved in museums so that later researchers could reexamine and reinterpret the evidence in light of new discoveries.

The second step is to construct a provisional explanation, or hypothesis, for the observations. A good hypothesis should generate predictions, and those predictions can be used to test the hypothesis. Case Study 3 presents an example of how Eugene Dubois tested the prediction made by Ernst Haeckel about the nature of human ancestors.

Getting It Wrong: Initially

There is good science and bad science. Good science does not mean coming up with right answers all of the time, but it does mean following a rigorous methodology. There are many reasons why errors are made. Scientists may be working with bad observations or incomplete information. They may be building on incorrect hypotheses or erroneous assumptions that are deeply embedded in their culture. Individuals may also allow pride and prejudices to color their thinking.

One of the more common complaints in paleoanthropology is the paucity of the fossil record and the claim that a problem can only be addressed with "more fossils." At present, there are about 350 sites with hominin remains that are not modern humans. Many more contain archaeological evidence but no remains. Some of these sites have produced hundreds of fossil bones and fragments, a couple have thousands, and many have a little as a single tooth. Despite this impressive collection, the record remains dismally incomplete and limited to places and times where fossils were preserved in the past and exposed in the present and where anthropologists have looked for them. Consider, for example, that a thousand specimens from a period of a million years in the Old World is still only one fossil per thousand years. In an evolutionary sense, a hominin species is not likely to change very much in a thousand or even in 10,000 years. However, that one specimen per thousand years can only represent one point geographically and only one part of one population on one of three continents. Anthropologists attempt to build evolutionary trees based on what evidence they have, but most of the known fossils may lie on dead side branches and the true human ancestors from certain time periods may not yet have been sampled.

It is little wonder, therefore, that instead of filling in gaps, new finds often may bring more questions than answers. There now exists a reasonable record from East Africa from 4.0 to 1.5 Ma ago and likewise from the Transvaal Valley in South Africa from about 3.0 to 1.5 Mya. Nonetheless, a new species of australopithecine was named from Ethiopia and a new member of *Homo* from South Africa, both in 2015. Many expect that more species exist in the collections that have not yet been recognized.

It is easy to misinterpret such a sparse and ambiguous fossil record. We count on more discoveries to help us, but the larger scientific community plays an essential role in identifying and correcting errors. The peer review process assesses the appropriateness and significance of new findings and interpretations before they are published, but scrutiny continues long after that. The standard path of a scientific claim is for scientists to review, replicate, and build upon the work of one another. Sometimes problems are only uncovered when new tools and methods become available; sometimes new fields of inquiry are inspired by hypotheses that don't seem right. When contradictions appear, it is incumbent upon scientists to resolve them, determining the cause and correcting errors.

A number of case studies illustrate that process. The notorious Piltdown hoax (Case Study 4) produced a fossil that misled anthropologists for 40 years before it was uncovered. Scientists must work within constraints, however, which include respect for data. The literature of those years reveals much about how researchers struggled to deal with an increasingly anomalous specimen. Case Study 7 involves a genuine fossil, *Ramapithecus*, wrongly assigned to a key role at the start of the hominin lineage. The invention of a new line of inquiry, molecular anthropology (Case Study 6), challenged that model and inspired a decade of research to resolve the contradiction. In Case Study 9, anthropologists wrestled with one of the most abstract of subjects, human nature, and inevitably interpreted the data through their cultural biases. A false start encouraged the development of a new field, taphonomy, to test claims about the behavior of our ancestors. Case Study 18 argues that the interpretations that take place after discovery may still be biased by our expectations and we must be open to alternative views.

The accompanying table is offered as a summary of themes in content and science to assist instructors in using these case studies within their curricula.

Table 1 Thematic outline of the case studies of this book

Cast study	Paleoanthropological issue	Taxon	Practice of science
1	Evolutionary theory	Life	Paradigm shift
2	Establishing prehistory	Paleolithic humans	Systematic data collection
3	Testing evolution	*Homo erectus*	Hypothesis testing
4	Recognizing and rejecting a hoax	Piltdown hoax	Constraints on scientific method; self-correction by the scientific community
5	Geological dating	*Paranthropus boisei*	Interdisciplinary collaboration; introduction of new technologies
6	Phylogeny of modern taxa	Living hominoids	Introduction of new technology; revising models for unexpected data
7	Relating extinct and living taxa	Miocene hominoids	Preconception bias; resolving competing hypotheses
8	Taphonomy	*Australopithecus*	Social construction; hypothesis testing
9	Anatomy of bipedalism	*Australopithecus afarensis*	Comparing competing hypotheses
10	Reconstructing stature; models for body form	Australopithecines	Cross-disciplinary applications; identifying appropriate analogies
11	Oldowan technology	Early *Homo* in East Africa	Experimentation
12	Diet and hunting	Early *Homo* in East Africa	Hypothesis testing
13	Paleoclimate	Early *Homo* in East Africa	Hypothesis testing
14	Postcranial evolution and endurance	Early *Homo* at Dmanisi	Constructing a model
15	Life history strategy, maturation	*Homo ergaster* at Nariokotome	Identifying appropriate analogies
16	Mosaic evolution	*Homo naledi* and others	Access to fossils

(continued)

Table 1 (continued)

Cast study	Paleoanthropological issue	Taxon	Practice of science
17	Island dwarfing	*Homo floresiensis*	Revising models for unexpected data
18	Reconstructions, recognizing humane behavior	*Homo neanderthalensis* at Shanidar	Projections of biases onto past hominins
19	mtDNA; modern human migrations	Modern humans	Introduction of new technology
20	Species relationships	Neanderthals and modern humans	Fitting models to unexpected data
21	Modern behavior	Early modern humans	Comparing competing hypotheses
22	Ancient DNA	Archaic and modern humans	Introduction of new technology
23	Ecological position	Hominins	Identifying appropriate analogies
24	Bipedalism	Hominins	Limits of scientific inquiry
25	Aquatic ape and waterside hypotheses	Hominins	Competing paradigms; umbrella hypotheses
26	Intelligent design	Life	Defining science

Getting It Right: Eventually

There are many sources of breakthroughs in science, and both revolutionary approaches and the slow patient accumulation of data are important. The examples in this volume note both. Often progress is made by the application of technologies and methods from other disciplines, such as geophysics (Case Study 5), molecular biology (Case Study 6), forensic sciences (Case Studies 10 and 15), and genomics (Case Studies 19 and 22). At other times, it is our ability to step back and take a newer perspective on years of basic studies that leads to new understandings, for example, about bipedalism (Case Studies 9 and 14), the paleoenvironment (Case Study 13), or revolutions in cultural behavior (Case Study 21). Another path to better insight is to ask new questions. Examples here examine early tools (Case Study 11) and evidence for hunting (Case Study 12). Occasionally, it is an unexpected discovery that demands to be noticed and forces us to reexamine what we thought we understood, such as a primitive species whose dead appear to have been deliberately deposited in a cave (Case Study 16), the enigmatic Hobbit (Case Study 17), unexpected old dates for modern fossils (Case Study 20), or genetic evidence for unknown hominin populations (Case Study 21).

The final case studies attempt to understand the limits of science. Anthropology tends to lose its objectivity when it explores human behavior and human nature. Our uniqueness as a species is more apparent than real (Case Study 23); perhaps it is most apparent in our ability to ask such questions. Some questions about the past are simply beyond resolution from direct scientific inquiry (Case Study 24) or lie outside the rules of science.

Science is a powerful tool. Its strength comes from its rigor and its rules. There are movements in our society that are unhappy with its findings and want to bend its rules to justify the outcomes they desire. The final case study (26) comes not from a scientific study but a legal one that reaffirms that our society recognizes natural science as a discrete and important exercise of the human mind.

I hope students can come away from a studying a fractious discipline that is fraught with subjective preconceptions and appreciate the positive role that science can play in bringing bias to light and establishing standards for recognizing more reliable truths.

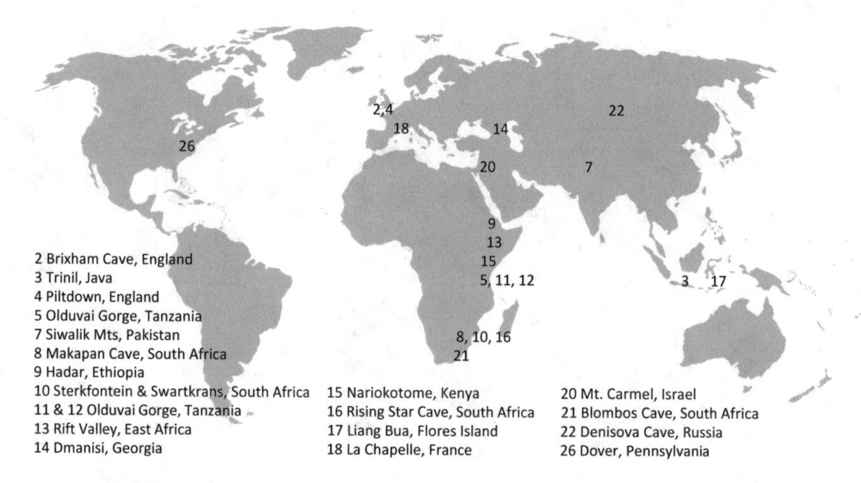

Fig. 3 Locations of sites discussed in the Case Studies in this book. Modified from https://commons.wikimedia.org/wiki/File:BlankMap-World-noborders.png#file with permission

Case Study 1. The Darwinian Paradigm: An Evolving World View

Abstract One of the most influential interpretations of the history and philosophy of science was that of Thomas Kuhn, whose book, *The Structure of Scientific Revolutions* (1962), introduced the term "paradigm" into popular vocabulary. In Kuhn's understanding of science, science constructs a world view, or paradigm, that shapes the way we view the world and conduct or pursuit of science. When major theories are discarded and replaced, we have rejected one set of assumptions for another and undergone a revolution in thought. The most significant "paradigm shift" that has taken place in the biological sciences was the Darwinian Revolution, which introduced not only evolutionary thinking, but also the scientific method.

Thomas Kuhn's *The Structure of Scientific Revolutions* is a now-classic perspective on how science "progresses." Major breakthroughs, he argues, occur when we move out of an existing paradigm into a new one. Although he does not rigorously define the term, Kuhn is largely responsible for introducing "paradigm" to the philosophy and history of science, and the term quickly moved into general use. In his usage, a paradigm is a broad theory, consistent with existing observations, that provides a worldview within which further observations, experiments, and hypotheses may be interpreted. A paradigm is constructed from certain postulates, or assumptions. The paradigm determines what questions can be asked and investigated and constrains the nature of possible answers. The pursuit of questions within the discipline is "normal science" and describes the activities of most researchers.

If the postulates are rewritten, the paradigm changes; however, since data are gathered and hypotheses constructed under the existing paradigm, it is very difficult to challenge and test those starting assumptions. Pressure to change a paradigm builds when anomalous observations accumulate that it has been unable to predict and explain. The possibility of change occurs only when a new paradigm is conceived that incorporates and explains existing observations and the anomalies. However, because this requires rejecting familiar assumptions, this is a difficult step. Shifting paradigms is so dissonant, a new paradigm is likely to attract mostly younger scientists less invested in the old one, and the community as a whole shifts gradually as the new generation replaces the older one.

J.H. Langdon, *The Science of Human Evolution*,
DOI 10.1007/978-3-319-41585-7_1

Certainly the most important paradigm shift in biology has been the acceptance of organic evolution. This was only one part of a shift in thinking associated with the rise of modern science.

The Pre-Darwinian Paradigm

The pre-Darwinian paradigm was built largely upon Aristotle's work, including *The History of Animals*, a volume from his encyclopedia. Like much of knowledge through the Middle Ages, biology was regarded as received wisdom, based on the writings of a few classical scholars with minimal additions. The modern scientific practice of verifying and adding to knowledge through observation was not an expected practice. Much more effort was spent aligning facts with theoretical and philosophical concepts to achieve a more complete understanding of the universe. A second unquestioned assumption was that the world was unchanged in any significant way since its beginning. To be fair, few humans witnessed significant changes in society, technology, patterns of living, customs, dress, language, or nature through their lifetimes until the modern era. They would have had little basis for thinking in terms of long-term linear change.

Aristotle, to his credit, used empirical observation, including dissections, to investigate zoology. He cataloged and classified a wide range of types, and discussed not only their anatomy, but also mating habits, behavior, and ecology. Modern zoologists have many corrections and additions to make, but this is a remarkable achievement for one person working in near isolation.

Aristotle's science was adapted into Medieval Christian thought. Merged with a literal acceptance of the Genesis account of creation and a belief than a perfect creation implies an effectively unchanging state of the universe, his understanding of nature became dogma. His ideas were not challenged simply because the paradigm did not recognize the possibility of changing them.

Aristotle's system of classification was based on shared characteristics, but its logic is less apparent today. For example, he divided animals first into those with blood and those without blood. The former group consisted of animals that lay eggs and those that bear live young. The latter contains four divisions: insects, nonshelled crustaceans (e.g., octopus), shelled crustaceans, and molluscs. This morphed over the next two thousand years into the *scala naturae*, or Great Chain of Being. In the Middle Ages, the *scala naturae* formed a continuous arrangement of objects from minerals at the bottom to God at the top, representing increasing complexity, vitality, and spirituality (Fig. 1). Aristotle's study was descriptive, not explanatory. It was consistent with his larger philosophical perspective of teleology—the world is the way it needs to be. Animals have traits because they need them and lack traits they do not need. Thus, even though Aristotle practiced empirical observation, his work did not particularly enjoin or encourage others to do so.

Natural philosophers of the past were thus able to describe species and place them in relation to others. In the process, humans were regarded at the center of earthly life (just below angels). Teleology could be used to explain the observed

Fig. 1 A simplified version of the *scala naturae* depicting the ladder of creation from rocks at the bottom, through plants, animals, humans, and angels. Source: Ramon Llull (1304)

adaptiveness of animals, particularly when placed in the context of a benevolent Creator. However, because it was descriptive, the field was not able to generate predictions. The teleological approach to adaptation, with the assumption of perfect creation, was tautological.

Anomalies

Kuhn's model anticipates that "normal science" operating within a paradigm will accumulate anomalous observations that cannot be explained by the original theories. Normal science in the pre-Darwinian paradigm would have been content with describing and classifying new species of organisms. However, Age of Discovery and the rise of empirical thinking in the Enlightenment produced a steady stream of

Table 1 Examples of anomalies accumulating within the pre-Darwinian paradigm that brought about a crisis and paradigm shift

Discoveries of new species (e.g., species from new continents and microscopic organisms)
Species did not fit existing categories (e.g., platypus and kiwi)
New variants challenged boundaries of species (e.g., moose and American bison)
Discoveries of extinct species
Fossil record showing directional change through time
Uniformitarianism showed great age of earth
Geographical clustering of related species
Inconsistent distribution of species groups
Presence of vestigial structures without function
Homologies of structures across species
Additional homologies appearing in embryological development

anomalies and patterns that the existing framework could not explain (Table 1). These are the conditions that lead to a paradigm shift.

The first problem was the flood of new organisms to come to the attention of naturalists. Each new exploration into Africa, Asia, the Americas, Australia, and islands around the world brought species never imagined into Europe (Fig. 1). Many of these new species did not fit into existing classification. How could the Aristotelian system handle the platypus, an egg-laying warm-blooded mammal, or the kiwi, a wingless burrowing bird? New discoveries also challenged the understanding of existing types of animals. Why was the American moose so different from European elk even though they were obviously related? Which was the true elk? Why were swans white in Europe, but black in Australia? The invention of the microscope opened up new realms, as well, of minute but complex animals as well as single-celled organisms.

Early studies of geology were interested in minerals of economic interest, but soon began to appreciate fossils for their ability to correlate strata across the countryside. The fossils were bones and shells of unknown animals; why had they gone extinct? Some naturalists thought this inconsistent with the idea of a perfect creation. At the same time, the principles of stratigraphy and uniformitarianism were evidence of a great age of the earth. The fossil record showed systematic linear change. The further back the strata reached in time, the more different the ancient species appeared. These ideas inspired visions of past worlds quite unlike the present.

Yet another pattern began to appear that did not fit expectations. Instead of being scattered across the earth, animal species differed in different parts of the world. The animals of South America were not the same as those of Asia or Africa, despite living in similar habitats. In some areas, such as Australia, they were markedly different. Nearly all mammals in Australia were marsupials, and more like one another than like mammals from any other place. At the same time, the marsupials had adaptations that resembled those of wolves or cats or badgers or grazing placental mammals. Many islands had unique species of birds found nowhere else. Why would a Creator have made different types for different regions?

b

ORNITHORHYNCHUS ANATINUS.

Fig. 2 The discovery of new continents in the fifteenth and sixteenth centuries brought new species to the attention of Western scientists that did not fit into the existing classification system, including (**a**) the kiwi and (**b**) the platypus. This was one of many anomalies that led to the Darwinian Revolution. Sources: (**a**) John Gerrard Keulemans, Ornithological Miscellany. Volume 1; (**b**) John Gould, *The mammals of Australia*. Volume 1

Aristotle noted homologous structures that could be compared among related species—organs, limbs, etc. Not only did later naturalists observe the extension of homologies to newly discovered species, but also they observed deeper patterns. For example, the skeleton of a bird wing is much more similar to the forelimbs of land animals in its internal structure and identification of individual bones than external comparison would suggest. Some of these homologous structures were nonfunctioning vestiges, such as the pelvis of a whale or hind limb bones in a snake. These flatly contradicted the expectations of a teleological model. Studies of embryonic development extended this pattern. The human embryo, like those of all mammals, temporarily has structures like those of gills in fishes.

Naturalists were seeking explanations, not merely descriptions; and Aristotle's understanding of life could not explain these patterns. However, the idea of change through time suggested by the geological strata laid the foundation for a new paradigm. Many naturalists began to work with the concept of evolution, most famously Georges Cuvier (1769–1832) and Jean-Baptiste Lamarck (1744–1829). Their ideas lacked a clear mechanism that could explain how organisms could change and new species could arise. They also fell short of a comprehensive theory that could explain all of the many newly perceived patterns outlined above. It was Charles Darwin (1809–1882) who provided the mechanism, natural selection, and the grand vision and systematic supporting evidence from around the world.

The Darwinian Paradigm

Darwin's theory of evolution through natural selection did lead to a paradigm shift throughout the life sciences. Among the unquestioned assumptions of the new paradigm are deep time, uniformitarianism, and prehistory. The earth is very old and we can extrapolate natural laws and processes back into this "deep time." Geological ages extend well before human existence and, importantly, well before any written records. Any attempt to understand what happened during the early periods must be inferred from the geological record.

Charles Lyell (1797–1875) is credited with stating the principles of uniformitarianism. His studies of geology revealed example of uplift of sections of rocks during earthquakes. If extrapolated back in time through successive events, they could explain great changes in the landscape, even including mountains. Likewise, the daily erosion due to water and wind and occasionally greater floods might account for the creation of valleys and canyons and the wearing down of mountains. Lyell generalized to argue that all of the earth's landforms could be understood by the same phenomena we can observe in our lifetimes. In other words, the processes of nature are uniform across time, and therefore past natural history is knowable. Uniformitarianism applies to natural laws, the processes that cause change, and the rate of change.

All the lines of evidence tell us that the earth and the life on it have been changing through time; thus evolution is one of the primary inferences of the paradigm. Other important arguments based on empirical evidence are that species change through time with some going extinct and new ones arising; that species are populations that inherently have variation; that all organisms share a common ancestor; and that biological classification, geographic distribution, homologies of structure, and embryonic development all reflect evolutionary history. These claims are contrary to Aristotle's view but are perfectly sensible in the new paradigm. Evolution has tremendous explanatory power. It allows us to place any new living or fossil species in a phylogenetic tree. It explains adaptiveness and, more interestingly, explains non-adaptive traits. It explains similarities between unrelated species through convergent evolution. It explains peculiarities in the geographic distribution of species and geographic clustering of related species. It provides an explanation for the fossil records.

An evolutionary perspective explains adaptation as well as the previous paradigm, but also provides a mechanism, natural selection, to tell us how it may have originated. It addresses the anomalies of the old paradigm by accounting for the diversity of species, their geographic distribution, and the change over time. It also allows us to make predictions, which can be used to test and refine our theories. It predicts that all life is similar in some ways because it shares a common origin. We have confirmed that down to the molecular level and we can use homologies and vestigial structures to reconstruct phylogenies. It predicted that as we understand the mechanism of inheritance we would also understand how novelties could appear. It predicted that the fossil record would reveal transitional form between groups of animals, and we now have abundant examples of that.

The paradigm shift did not occur overnight. The intellectual revolution that began in England in the middle of the nineteenth century has been extensively documented and analyzed. Darwin's most enthusiastic converts and promoters tended to be younger scientists, such as Joseph Hooker and Thomas Henry Huxley, who looked up to Darwin as a mentor. Established scientists, heavily invested in the older paradigm, were more likely to be skeptical. Charles Lyell, whose books on geology inspired Darwin accepted evolution eventually, but some leading voices, including the anatomist Robert Owen and the Swiss-American geologist Louis Agassiz never did. This pattern, in which the rising generation is more open to new ideas, is familiar and widespread. Some of the older ideas were deeply embedded in cultural consciousness and intuition. The *scala naturae* appears in Haeckel's evolutionary scheme, though now it was a reconstruction of our past history instead of a description of contemporary rankings (Case Study 3). Both Haeckel and Alfred Russell Wallace, an independent discoverer of the concept of natural selection, held onto somewhat mystical notions of a providence guiding evolution to higher levels of perfection (i.e., humans). Although modern biologists have made efforts to distance themselves from these ideas, they persist in popular perceptions of evolution.

The Darwinian paradigm still makes many people uncomfortable because it contradicts assumptions of competing paradigms, especially those concerning our own place in nature. Whereas older views placed humans as the focus and purpose of creation, in the new perspective the question of purpose has no meaning. The old paradigm made humans superior to other species; in Darwinism there is no basis for claiming superiority of any species over another. Aristotle classified discrete species; Darwin recognized the fluidity of species over time. The old model emphasized purpose, morality, and relationship to God; the new model strives to understand adaptation, change, and organic relationships. In Kuhn's terminology, the two models are incommensurate. Accepting one does not make the other wrong, but nearly inconceivable.

Biologists now conduct normal science under the Darwinian paradigm. They ask questions about adaptiveness, construct phylogenetic trees, and attempt to reconstruct evolutionary history. Are anomalies accumulating that cannot be explained under the new paradigm? Very likely, but they are mostly hidden in the category of questions and observations we do not understand yet. Is it possible that the Darwinian paradigm is "correct" and describes nature so well that we will never need another paradigm shift? Kuhn speculated about the possibility of permanent "normalcy" in which further shifts are unnecessary, but then rejected it as most unlikely.

The Darwinian Revolution may be understood as a logical extension of the development of modern science. During the Enlightenment, a view of the universe emerged with the conviction that the laws of nature were comprehensible through empirical investigation. Experimentation, observation, and theorizing spread from one area of science to another, following the emerging rules of the scientific method. As knowledge and technology increased, branches of science that we now recognize as separate disciplines diverged. The Copernican Revolution, Kuhn's model paradigm shift, may be understood as the beginnings of modern astronomy, to be followed by the emergence of physics, geology, chemistry, biology, medicine, and other scientific fields. Within these fields, theories are constantly being advanced, tested, and sometimes accepted. We may understand these as small paradigm shifts. However, we cannot expect new revolutions on the scale of Copernicus, Newton, and Darwin, because the greatest revolution, the advent of modern science, has already occurred.

Questions for Discussion

Q1: In what way does a scientific paradigm, or its starting assumptions, constrain the questions one can ask? Give examples.
Q2: What does it mean for two ideas to be incommensurate?
Q3: From what observations might the idea of the "*scala naturae*" have arisen?
Q4: What are the assumptions that underlie and define the modern paradigm of biology? Can they be tested?
Q5: Darwin's model was overtaken by the Modern Synthesis. Did that constitute anoehr paradigm shift?

Additional Reading

Lyell C (1998) Principles of geology (abridged). Penguin, New York
Mayr E (1985) The growth of biological thought: diversity, evolution and inheritance. Belknap, Cambridge
Kuhn T (1962) The structure of scientific revolutions. University of Chicago Press, Chicago

Case Study 2. Proving Prehistory: William Pengelly and Scientific Excavation

Abstract Science is empirical, based on sensory observations. Those observations must be repeated or repeatable and objective. During the eighteenth and nineteenth centuries, the study of natural history developed from a hobby of the educated elite, often reporting isolated or unsystematic observations, to a profession with careful methodologies. William Pengelly, the subject of this chapter, developed a system of careful excavation and recording of finds at prehistoric sites that is still in use today.

By the early 1800s, through the work of such people as Nicolaus Steno (1638–1686), James Hutton (1726–1797), and John Playfair (1748–1819), it was widely recognized that the earth's geological formations are the products of natural processes. Charles Lyell (1797–1875) assembled numerous observations from around the world to argue that that volcanism—including volcanic activity and earthquakes—could build up the land, whereas the action of water eroded it away. Over long periods of time, these familiar processes could account for immense changes in landforms. The concept became codified as uniformitarianism, the understanding that the processes and laws that acted in the past were the same that we observe in the present. This was one part of a more profound revolution in worldview that emerged in the eighteenth and early nineteenth centuries—the discovery of a prehistory before humans when other kinds of life inhabited the earth.

Into this geological deep time, naturalists learned to place fossil animals in predictable sequences, due in part to the work of William Smith (1769–1839). Smith, a surveyor, became aware that sedimentary rocks were laid down in a specific pattern that was recognizable across large swaths of England. Each of these layers, or strata, could be identified by the presence of distinctive assemblages of fossils. Smith documented the strata and began a catalog of fossils, particularly noting common and distinctive species (index fossils) that would identify a layer with the greatest certainty. Before he had completed his work for England, others had already begun applying his approach in France. Soon it was possible to correlate rocks across Europe to create the beginning of the geological time scale (Table 1).

J.H. Langdon, *The Science of Human Evolution*,
DOI 10.1007/978-3-319-41585-7_2

Table 1 Geological time was worked out in the nineteenth century on the basis of successive changes in the fossil record. While layers of sediments and fossils could be assigned relative dates, naturalists could not assign absolute dates until the mid-twentieth century

Era	Period	Epoch	Major events
Cenozoic	Quaternary	Pleistocene	Ice ages; Modern genera appear; *Homo sapiens* emerges
	Triassic	Pliocene	First *Homo*
		Miocene	Hominoids radiate; hominins diverge; grasslands spread
		Oligocene	Anthropoids diversity in Africa
		Eocene	First anthropoid primates
		Paleocene	Mammals dominate; modern orders appear; first primates
Mesozoic	Cretaceous		First flowering plants; dinosaurs, pterosaurs, large marine reptiles extinct at end
	Jurassic		Dinosaurs dominant; first birds; marine reptiles dominate oceans
	Triassic		First dinosaurs, pterosaurs; first mammals
Paleozoic	Permian		First vascular plants; synapsid reptiles dominate; greatest extinction event at end
	Carboniferous		First reptiles
	Devonian		Animals invade land; first insects, first amphibians
	Silurian		First jawed fishes
	Ordovician		First fishes; trilobites common
	Cambrian		Diversification of animals, rise of modern phyla; first vertebrates

The majority of these fossils were neither from any known living animals, nor from any described by classical writers. Human artifacts appeared only in the more recent layers. Debate arose around the question of whether or not humans coexisted with extinct animals. Did ancient writings encompass the full antiquity of human existence, calculated from Biblical genealogies to only the last 6000 years or so, or were humans present in the time before the written record—literally in "prehistory"?

Early archaeologists in the 1800s occasionally reported finding human remains and stone tools intermingled with fossils of extinct animals in caves in France, England, and Belgium. Because of the importance of the questions at stake and the unsystematic methods of digging for artifacts, the scientific establishment maintained a skeptical reluctance to accept such claims at face value. In order to resolve this debate, it would be necessary to find the bones of extinct animals and evidence of humans intermingled in a context that had not been disturbed, and to do so in the presence of expert witnesses.

Brixham Cave

The opportunity to settle the debate arose in 1858 at Brixham Cave near the town of Torquay in southwestern England. Workers at a quarry broke through a rock wall into a previously unknown cavern, whose natural entrance had been blocked long ago. Parts of the floor of the cave were sealed over by flowstone left behind by evaporating water. Bones and fragments of antlers from extinct animals on the surface and embedded in the flowstone gave evidence of its antiquity. Because the older sediments were sealed off from any later disturbance, this cave would prove to be a good place to understand how such deposits were created and, incidentally, to look for evidence of human antiquity.

The discovery of the cave came to the attention of William Pengelly, a local schoolmaster, experienced geologist, and member of the Torquay Natural History Society. The society agreed to excavate the cave, but needed to raise money to pay its owner. Since Pengelly was also a member of the prestigious Geological Society of London, he sought and obtained support from that body as well, thus attracting the attention of the international community. The undisturbed deposits offered the possibility of investigating the sequence of animals that inhabited England during the Pliocene and Pleistocene and of refining stratigraphy during that time. As the potential importance of the excavation became apparent, the London scientists paid closer attention to the cave and urged Pengelly to excavate with meticulous care.

Pengelly invented a systematic method of investigation. Many of his contemporary prehistorians dug holes more or less randomly in search of fossils, destroying context and evidence. Once their bones had been unearthed, it may no longer have been clear whether they had originally lain at different levels and come from different ages. Pengelly directed his workmen to remove sediments carefully in layers. Each find, whether a bone or a stone tool, was exposed in place and its position recorded, both in distance from the entrance and in depth, before it was collected. Pengelly described his methods in this way:

> We make a vertical section down through the deposits, say at ten feet from the entrance, at right angles to a datum line drawn horizontally from a point at the entrance to another at the back of the first chamber, in the direction, as it happens, of W. 5° N. magnetic. We draw a line at right angles to the datum at eleven feet from the entrance so as to define or mark off a new "parallel" a foot wide. Along this entire belt or parallel we take off the black mould from side to side of the chamber, and examine it carefully by candlelight *in situ*. Another man takes it then to the door and re-examines it carefully by daylight. All the objects found in it are put into a box, which is numbered, and a label is put in with them. We proceed with the stalagmite [i.e., flowstone] in like fashion; we then come to the cave earth, where we are still more particular. We take a piece simply a yard in length and a foot in depth—in short, a parallelpiped a yard long and a foot square in the section and termed a "yard." We examine that in like manner, and what we get is put into a box, and so on yard after yard and level after level to a depth of four feet below the granular stalagmite. All the boxes thus filled during the course of a day are sent to my house in the evening (Pengelly 1876).

Thus with the scientific establishment watching over his shoulder, any discovery of prehistoric humans would be witnessed and carefully documented.

Over the course of two seasons, Pengelly recovered 1621 bones of at least 20 different mammalian species. Most of these represented animals long extinct from the British Isles, including mammoth, wooly rhinoceros, cave lion, hyena, and cave bear. They were readily recognized as belonging to the Pleistocene Epoch, the time of the Ice Ages. No human bones were uncovered, but 36 flint tools and flakes were found between 6 and 18 ft deep in the cave floor, deeper than many of the animal bones.

Although initially skeptical about the nature of the "tools," some members of the Royal Society became convinced of their human origin once they had a chance to see the tools for themselves. The symmetry, complexity of manufacture, and similarity to tools known from other sites left no doubt that they were genuine. Inspired by this evidence of human prehistory, geologist Hugh Falconer and other members of the Society visited excavations on the continent where claims of similar association of humans with Pleistocene animals had been viewed skeptically. They soon confirmed the antiquity of human presence at Manchecourt, where Boucher de Perthes was currently excavating, and at other sites in France—Moulin-Quignon, St. Roch, and St. Acheul—and at Grotta di Maccagnone, in Italy. The existence of Pleistocene humans finally had the approval of the scientific establishment in England (Fig. 1).

Fig. 1 Prehistoric stone tools from Southeastern England (Gough's Cavern, Cheddar) not far from Brixham's Cave. Source: Geological Society of London 1845. Source: Haeckel, Ernst. *The Evolution of Man: A Popular Exposition of the Principal Points of Human Ontogeny and Phylogeny*. New York: Appleton & Co., 1897

Pengelly had the opportunity to apply and further refine his techniques in the much larger Kent's Cavern nearby. Previous digging there had already uncovered tools and animal bones, but those efforts had been relatively unsystematic and their findings were not accepted by the scientists in London. Again, flowstone covering areas of the cave floor guaranteed that underlying deposits had been undisturbed. Once more the sediment was removed in blocks three feet wide, one foot across, and one foot deep. Pengelly plotted the positions of objects found here in three dimensions. Over a period of 12 years (1868–1880), Pengelly revealed a complex stratified sequence of deposits. Those excavations and more that have continued to the present have uncovered over 100,000 bones and artifacts. The oldest tools go back 450,000 years. Some human remains, including a partial jaw now attributed to a Neanderthal, have also been unearthed.

The Principle of Superposition and Relative Dating

The inference that tools and bones found side by side had coexisted in the past may seem self-evident today to anyone familiar with archaeology, but it requires certain assumptions. Sedimentary rocks are generally found in distinguishable horizon strata. It is inferred that objects in the same layer were deposited within the same span of time, that the bottom layers are the oldest, and that higher strata were put down later. Such inferences have been codified as the Principle of Superposition. Although the principle may appear self-evident, that has not always been the understanding. In the eighteenth century, a concept of geology known as catastrophism competed to explain the world. Catastrophists believed that a few violent world-changing events, such as a global flood, could have created the landforms we observe in a short period of time, as in the biblical week of creation. In such a model, all the strata would be effectively contemporary in their formation, but should also contain objects mixed together that had originated from different previous time periods. Science has rejected catastrophism in favor of uniformitarianism.

The significance of the Principle of Superposition, first formulated by Steno in the 1660s, goes beyond the argument that sediment is created over a period of time. It allows us to establish systems of relative dating. Relative dating has tremendous potential for helping us to understand the past even if we do not know any absolute dates. We like to know exactly how many years old a fossil or event is, but more often we are only able to place it into a sequence where it might be compared with older or younger fossils. However, if some of those species exist only for defined periods of time, such as the large mammals of Ice Age Europe, then the presence of such fossils anywhere help us to establish relative chronologies. Important fossils, which were widespread geographically but only lived a brief time, are known as index fossils. Distinctive fossils, sediments, layers of volcanic ash, coins, styles of stone tools or pots, or any other identifiable phenomenon can serve as a time marker. The more restricted in time it is, the more useful it can be in establishing chronology. If each column of geological strata represents a time sequence, and if each column contains fossils or minerals that permit us to relate it to the other column in another location, then we can establish a regional, if not global, table of geological time.

Often strata are not put down in flat layers as simply as we would like them to be. For example, sediments settling in uneven surfaces, such a riverbed or a cave, naturally conform to that shape, so that older deposits may now lie beside the newer ones. Pengelly's systematic approach to excavation helps us to make sense of complex deposits. By mapping the finds in three dimensions, and also describing variations in the soil surrounding them, it is possible to record how they are clustered together and how they relate to the different layers of sediments.

For these reasons we must understand that a fossil or artifact taken out of context has limited value. Where it was found and at what depth are vital clues for understanding it. Professional archaeologists and paleontologists know that such information must be carefully recorded and preserved, and careless collectors and looters destroy valuable scientific information.

Questions for Discussion

Q1: The idea of long-term historical change came slowly to people. What evidence of social and cultural change have you observed in your lifetime? What would have been available to people in the Middle Ages?

Q2: How might the idea of deep time change people's perspectives on themselves and the world in which they live?

Q3: In the nineteenth century, "archaeologists" often dug just to see what they could find, if not for more mercenary aims. Pengelly tried to answer a specific question about the change in the animal community in the Pleistocene. What difference does it make if the excavator has a specific question in mind or not?

Q4: Why did Pengelly think it might be important to record the exact position of each find in the caves?

Q5: Describe in your own words Pengelly's method of recording his discoveries to prove that the tools were as old as some of the bones of extinct animals.

Q6: Archaeologists and paleontologists destroy context and information when they excavate. How can they best prevent the loss of that knowledge?

Q7: What happens if we do not assume uniformitarianism and consider a past (or future) in which natural laws and scientific constants may have been different? How would that affect the conduct of science?

Additional Reading

Goodrum MR (2004) Prolegomenon to a history of paleoanthropology: the study of human origins as a scientific enterprise. Part 2. Eighteenth to the twentieth century. Evol Anthropol 13(6): 224–233

Grayson DK (1983) The establishment of human antiquity. Academic, New York

McFarlane DA, Lundberg J (2005) The 19th century excavation of Kent's cavern, England. J Cave Karst Stud 67(1):39–47
Pengelly W (1876) Kent's Cavern: its testimony to the antiquity of man. W Collins, London
Prestwich J (1871–1872) Report on the exploration of Brixham Cave, conducted by a committee of the Geological Society, and under the immediate superintendence and record of William Pengelly, Esq., F.R.S., aided by a local committee; with descriptions of the organic remains by G. Busk, Esq., F.R.S., and of the flint implements by John Evans, Esq., F.R.S. Proc R Soc Lond 20:514–524
Van Riper AB (1993) Men among the mammoths: Victorian science and the discovery of prehistory. Univ Chicago Press, Chicago

Case Study 3. Testing Predictions: Eugene Dubois and the Missing Link

Abstract As the implications of Darwin's theories on human evolution were absorbed by the scientific community, interest grew to understand the biological nature of our ancestors. In Germany, Ernst Haeckel constructed a theoretical model of our history all the way back to single-celled organisms. Each of his 22 stages was a link in an evolutionary chain. Some ancestors were reasonably represented by living species, others were "missing." Among the missing links were the last two before humans. Haeckel named these man-like apes and ape-like men and described their essential characteristics. However, hypotheses that have not been tested are only informed speculation. We test hypotheses by making predictions and seeing whether those are fulfilled. Haeckel's model inspired Eugene Dubois, to go to the far side of the globe in search of the fossils to fill his gaps. In the following case study, Dubois ostensibly tested a vague theoretical abstraction; but what was really at stake is the hypothesis that humans evolved.

Reinterpreting the *Scala Naturae*

From the time of Aristotle, naturalists searching for a way to organize information about living organisms arranged animals on a linear continuum from simplest to most complex—the *scala naturae*, or the great chain of being. In the Middle Age spiritual beings—God and various ranks of angels were place at the top of the scale. At the bottom were inanimate objects, minerals. In between stretched the known plants and animals with humans at the top of the scale. Of course, living organisms do neither support a linear arrangement, nor is there a smooth continuum. It is self-evident to the casual observer that some animals (e.g., fishes or mammals) form clusters containing an equivalent level of complexity and that there are gaps between the clusters. Nonetheless, even in an evolutionary tree it is possible to trace a direct line of descent from any ancestor to us, conveniently ignoring the branches. Thus, the concept of the *scala naturae* was easily absorbed into evolutionary thought even though it perpetuates serious misconceptions by suggesting that we are descended from living species, such as chimpanzees. If this

J.H. Langdon, *The Science of Human Evolution*,
DOI 10.1007/978-3-319-41585-7_3

were true, we would have to assume that our ancestors had populations that, unlike us, simply stopped evolving. German biologist Ernst Haeckel showed his annoyance at this error when he wrote

> This opinion, in fact, has never been maintained by thoughtful adherents of the Theory of Descent, but it has been assigned to them by their thoughtless opponents. The Ape-like progenitors of the Human Race are long since extinct. We may possibly still find their fossil bones in the tertiary rocks of southern Asia or Africa.

(Despite this clear answer, opponents of evolution continue to raise this misunderstanding as an objection, either from ignorance or deceit.)

Haeckel proposed a model that incorporates a linear sequence (Fig. 1). He argued that each ancestral stage was represented at some point in the embryological development of the individual. This concept, captured in the English expression "ontogeny recapitulates phylogeny," enabled him to predict the characteristics of the "missing links." He characterized the 21st stage (between "Man-like Apes" and humans) as "Ape-like man (Pithecanthropi)," which he described as follows:

> Although the preceding ancestral stage is already so nearly akin to genuine Men that we scarcely require to assume an intermediate connecting stage, still we can look upon the speechless Primaeval Men (Alali) as this intermediate link. These Ape-men, or Pithecanthropi, very probably existed toward the end of the Tertiary period. They originated out of the Man-like apes, or Anthropoides, by becoming completely habituated to an upright walk, and by the corresponding differentiation of both pairs of legs. The fore hand of the Anthropoides became the human hand, their hinder hand became a foot for walking. Although these Ape-like Men must not merely by the external formation of their bodies, but also by their internal mental development, have been much more akin to real Men than the Man-like apes could have been, yet they did not possess the real and chief characteristic of man, namely, the articulate human language of words, the development of a higher consciousness, and the formation of ideas. The certain proof that such Primaeval Men without the power of speech, or Ape-Like Men, must have preceded men possessing speech, is the result arrived at by an inquiring mind from comparative philology (from the 'comparative anatomy' of language), and especially from the history of the development of language in every child ('glottal ontogenesis') as well as in every nation ('glottal phylogenesis'). (Haeckel 1876 vol. 2: 264)

Of the two characteristics that Haeckel singled out to define true humans—bipedal walking and speech—this putative ancestor possessed the first but not the second. He assigned this hypothetical creature a scientific name, "Pithecanthropus alalus," meaning "ape-man without speech."

From Theory to Fossils

Haeckel's exercise would have remained speculation had it not inspired a young Dutch doctor, Eugene Dubois, to attempt to find pithecanthropus. The problem was where. In *The Descent of Man*, Charles Darwin had famously suggested Africa as our biological homeland, probably because he favored the linkage to chimpanzees and gorillas. (However, he also added: "but it is useless to speculate on this subject.") Haeckel himself favored South Asia, or possibly a hypothetical lost continent

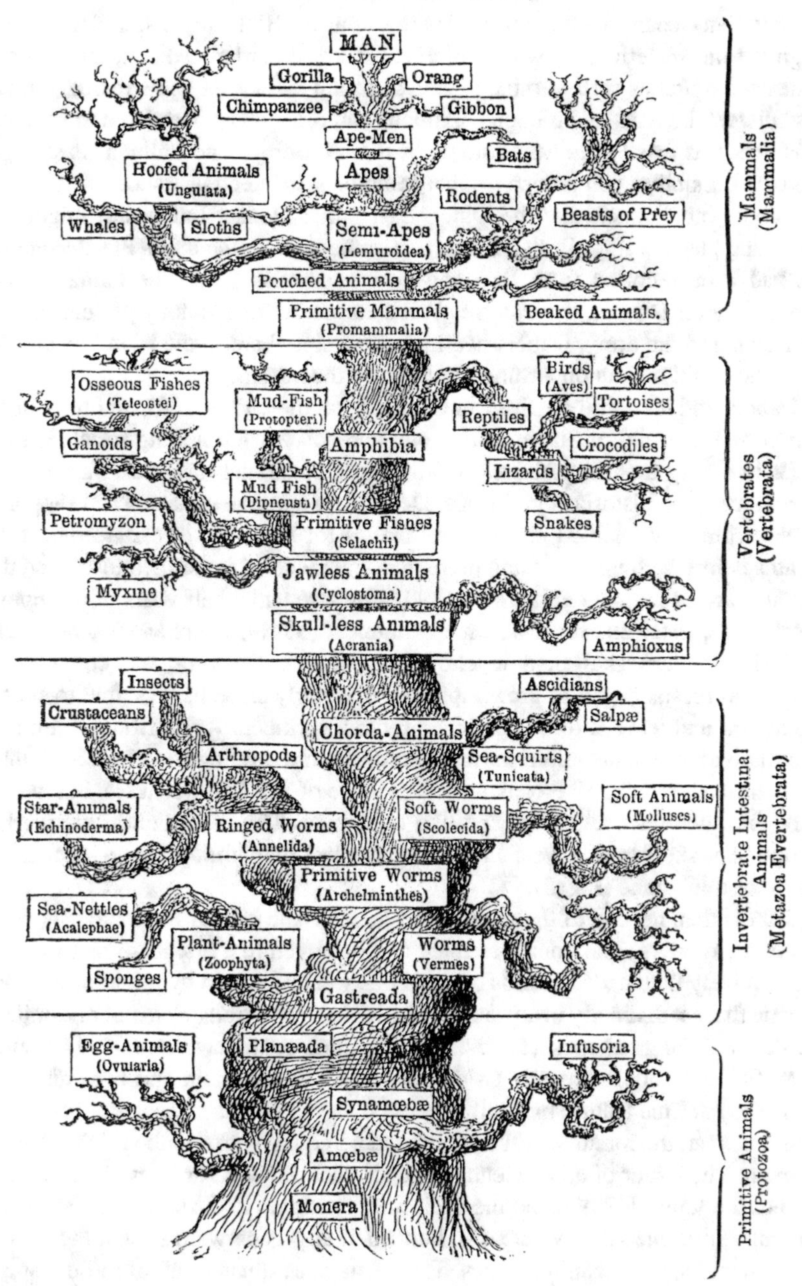

Fig. 1 Haeckel's phylogenetic tree with the stages of evolution links leading to humans. Originally published in *The Evolution of Man* (1897). https://commons.wikimedia.org/wiki/Ernst_Haeckel#/media/File:Pedigree_of_Man_English.jpg. Source: Haeckel, Ernst. *The Evolution of Man: A Popular Exposition of the Principal Points of Human Ontogeny and Phylogeny*. New York: Appleton & Co., 1897

in the Indian Ocean called Lemuria. There was also a Biblically inspired tradition of origins from western Asia, which, in the absence of hard evidence to the contrary, influenced attitudes well into the twentieth century. Haeckel had stressed similarities between humans and gibbons. Gibbons and orangutans both lived in southeast Asia, whereas fossil apes were known from both Europe and India. This strengthened the possibility that human origins could be discovered in Asia.

Dubois rationalized that the Dutch East Indies (today, Indonesia) might be a promising place to start. Both gibbons and orangutans reside there. Pleistocene fossils had been reported from Java and showed similarities to the fauna of Asia. Fortunately for Dubois, the Dutch East Indies was a Dutch colony. By enrolling as a physician in the army, Dubois had himself posted there to the island of Sumatra where he could find an opportunity to look for the missing link.

Dubois and his family arrived in the Indies at the end of 1887 and took up his station on the island of Sumatra. He soon appreciated that finding fossils is much easier in theory than in practice. Dubois expected to look in caves, based on the experience of prehistorians in Europe. Despite the approval and material support of his superiors that enabled him to hire natives for the work of excavating, it was months before he found anything more than fragments of bones and teeth, and then still no trace of human relatives. In 1890, after two and a half years of disappointment, he requested and received a reappointment to Java, where an ancient but anatomically modern skull had recently been found. On Java, the collection of nonhuman fossils began to grow rapidly, particularly since he was able to start an excavation and leave it to be continued by his workmen. Aside from a minimally informative piece of mandible, his first hominin find was a single tooth, in September, 1891, from an eroding riverbank near the village of Trinil. It was large, but its form could be interpreted either as ape-like or human. The next month, his workers recovered a skull cap. A year later came his last hominin specimen, a complete femur, or thigh bone (Fig. 2).

Dubois then needed to figure out what he had. The cranial fragment was primitive in many ways. The bone was thick and belonged to a braincase that was long and relatively flattened. The tops of the orbits were protected by pronounced ridges of bone that sat distinctly in advance of the brain. These traits were not dissimilar to the skulls of the great apes. However, the cranium would have contained a brain of about 940 cm^3, approximately twice the size of an ape brain and two-thirds of a human brain. This feature by itself suggested an intermediate species.

In contrast, the femur was fully modern and equal in size to the thigh bone of a tall man. The femur of an ape tends to be short and robust for body size, while the head of the femur is small and the neck sharply angled to the shaft. The shaft rises vertically from the knee, whereas in humans it angles outward so that the hips are more widely spaced than the knees. The fossil was human in all of these respects. (It also possessed an interesting pathological growth below the neck, probably due to calcification of soft tissues following an injury and infection.)

Together these two bones and the tooth, which Dubois assumed all came from the same animal, indicated a modern upright body paired with a primitive head and brain. His interpretation resembled the combination predicted by Haeckel.

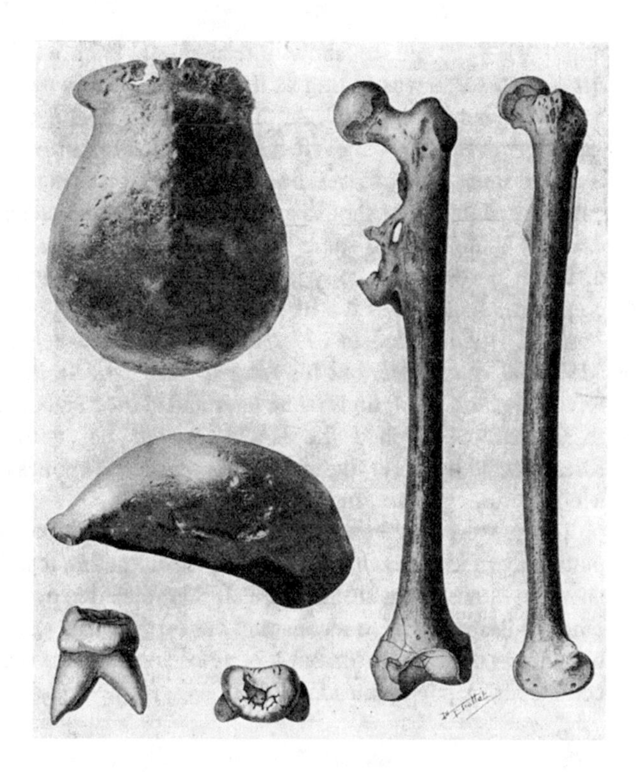

Fig. 2 The original *Pithecanthropus erectus* finds illustrated by Dubois (1892)

In acknowledgement, Dubois named the new creature *Pithecanthropus erectus*, "upright ape-man." In doing so, he recognized that this specimen lay on an evolutionary pathway between apes and humans, the first such fossil known.

With the benefit of hindsight and later discoveries, we can nod approvingly at Dubois' analysis. In the context of science of his day, his assessment is most reasonable, yet it was attacked mercilessly by scientists back in Europe. Critics claimed the femur looked human because it was human, and the skull must be from an ape. They questioned his interpretation of the sediments. They said the cranium was actually human and this was not a missing link. They said *Pithecanthropus* was an interesting creature, but not particularly related to us. They came up with every reasonable alternative interpretation.

This is the way science works. "Extraordinary claims require extraordinary evidence" (attributed to Marcello Truzzi). Certainly Dubois was making an extraordinary claim, and from halfway around the world he could not present the physical evidence to his critics. It is normal and essential that scientists explore all simpler explanations before accepting something revolutionary. They want to observe the evidence for themselves. It is possible, though unlikely, that a specimen could be pathological or otherwise so atypical as to disguise its identity. Acceptance should be cautious, as cruel as that may appear to the scientist trying to convince others.

One aspect of the fossils that undermined credibility was the unexpected combination of features. A missing link between apes and humans intuitively might be expected to show intermediate features throughout the skeleton. *Pithecanthropus* was telling us that the lower limb had achieved modern human form long before the head. At the turn of the century, many British anthropologists, in particular, believed that the expended brain was the defining human trait and must have evolved first. Of course, there is neither a reason why different parts of the body should evolve in a certain sequence, nor should we expect them to change at the same rate. When we observe such disparate patterns in different parts of the body, we use the term mosaic evolution, and it is quite common. Nonetheless it still often surprises us.

What Dubois need in order to convince the scientific community would be additional specimens, but his term in Java was limited and he was not able to repeat his luck. Such finds came later and at the hands of others. Although several fossil-seekers searched the island, it was not until 1931 that G.H.R. von Koenigswald began to find more pithecanthropus bones at the sites of Ngandong, Modjokerto, and Sangiran.

Today, we place these specimens and others in the species *Homo erectus*. By putting them in our own genus *Homo*, we are noting that they not halfway between apes and humans, but are far more closely related to ourselves. We now have many other "missing links" to reconstruct our evolutionary pathway. *H. erectus* is probably not our direct ancestor, but it is a close cousin. It is more likely descended from contemporary populations in Africa.

Dubois' Luck

On the surface, it appears that Dubois was unbelievably fortunate to have found a hominin fossil. If we were to dig naively at any site in the world chosen at random, the odds against making a similar discovery are astronomical. On closer look, Dubois had more insight than that with which he is generally credited. Although we now know that the early stages of human evolution occurred in a far distant continent from where he was working, looking in a region where apes exist today was not illogical. Moreover, humans have spread across the face of the globe, so there are few places without some traces of people however ancient or recent. Dubois and his contemporaries had only a hazy understanding of the geological time frame for human evolution; but he knew that humans in Europe coexisted with Pleistocene animals. The presence of Pleistocene fossils in Java told him the deposits may be of an appropriate age. It is a commonplace assumption today that tropical forests are poor places to find fossils because the acidic soils destroy bone quickly and because vegetation covering the ground makes it hard to prospect. However, Dubois knew fossils were being found in Indonesia, and he concentrated his efforts first and unsuccessfully in caves, and later along a river where strata were being eroded and exposed. Finally he must be credited for his persistence.

A modern field paleontologist would have many more advantages. We have a better, though still imperfect, idea of what hominins were living in a given part of the world at any one time. Geologists have mapped great areas of the earth's surface, so there is less guesswork about where rocks of a given age might be exposed. This information increases the odds that an expedition will be successful; but luck still plays a role. Few academics today have the luxury to afford years of prospecting without a significant discovery.

Dubois did have one advantage over modern researchers. He thought he was looking for *the* missing link to fill the gap between ancient apes and modern humans. So little we knew at that time that nearly any fossil hominin he found would fit that description. He discovered *a* missing link; and as soon as he named "*Pithecanthropus erectus*," he created two missing links—one earlier in time and one later.

Questions for Discussion

Q1: How did Haeckel's evolutionary sequence differ from the Medieval notion of the *scala naturae*?

Q2: Haeckel predicted the nature of intermediate species, particularly of the man-apes and ape-men. These predictions can be viewed at two levels. As a test of evolution itself, were the predictions falsifiable? As a test of a specific model of evolution and our ancestry, were the predictions falsifiable?

Q3: Haeckel's prediction appeared to be validated by Dubois' discovery. Was Haeckel insightful or lucky? What other forms might the missing link have taken? Is one of these more probable than others?

Q4: Why do new fossil discoveries rarely fill a gap between species? What would it take for us to be certain we have a complete evolutionary sequence, and why is that unlikely?

Additional Reading

Dubois E (1898) *Pithecanthropus erectus*; a form from the ancestral stock of mankind. Annu Rep Board Regents Smithson Inst 1898:445–459

Haeckel E (1876) The history of creation. Transl. E Ray Lankester. Henry S. King, London (Quotation from Vol. 2, p. 264.)

Shipman P (2001) The man who found the missing link. Harvard University Press, Cambridge

Case Study 4. Self-Correcting Science: The Piltdown Forgery

Abstract The scientific community plays an important role in critiquing hypotheses and building consensus and serves as a check to the influence of individual and societal prejudices. Although this system of challenge and debate looks chaotic to outsiders, it is essential for maintaining rigor and objectivity in science. The greatest scientific hoax of all time provides an excellent case to observe how science corrects an error. The Piltdown forgery succeeded for 40 years not because it was brilliant or the anthropological community was blinded by prejudices, but because scientists are bound by rules of evidence and because extrinsic factors in this case legitimately confounded the interpretation that evidence. One might say there was a fatal combination of bad luck and a naive reluctance to imagine malice. Skeptics appeared from the beginning and skepticism steadily grew, but the fossils could not be dismissed without resolving the anomalies.

The Piltdown Forgery

In February 1912, amateur geologist Charles Dawson wrote to Arthur Smith Woodward at the British Museum to inform him of the discovery of fragments of a robust human skull. Further excavations that year yielded more bones, including part of a decidedly ape-like mandible. Nonhuman fossils and flint tools accompanying the skull suggested a Pliocene age. Woodward and his colleagues reconstructed a human-like cranium from these pieces of bone, and put on it a very primitive lower jaw, naming it "*Eoanthropus dawsoni*." Thus began 40 years of debate over "the first Englishman" (Figs. 1 and 2).

Additional discoveries followed, including an unusual but long canine, an elephant-bone tool in the shape of a cricket bat, and other fossils and artifacts. A number of investigators remained skeptical until Dawson informed Woodward of a second discovery at a nearby site. This find, called "Piltdown II," consisted only of smaller skull fragments and a tooth, but the similar pattern of thick cranial bone plus ape-like dentition. Having an apparent second specimen confirmed the mosaic

J.H. Langdon, *The Science of Human Evolution*,
DOI 10.1007/978-3-319-41585-7_4

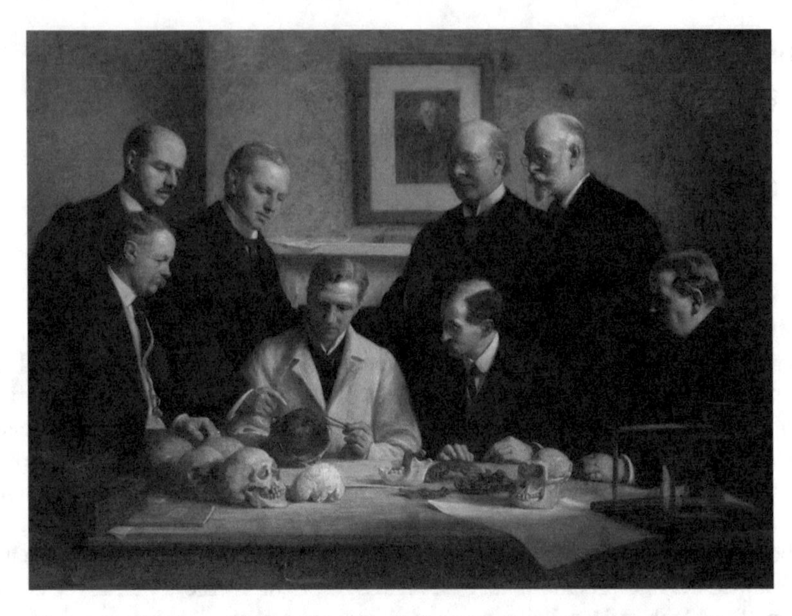

Fig. 1 The Piltdown scientists (painted in 1915). Charles Dawson stands in front of the framed picture of Charles Darwin

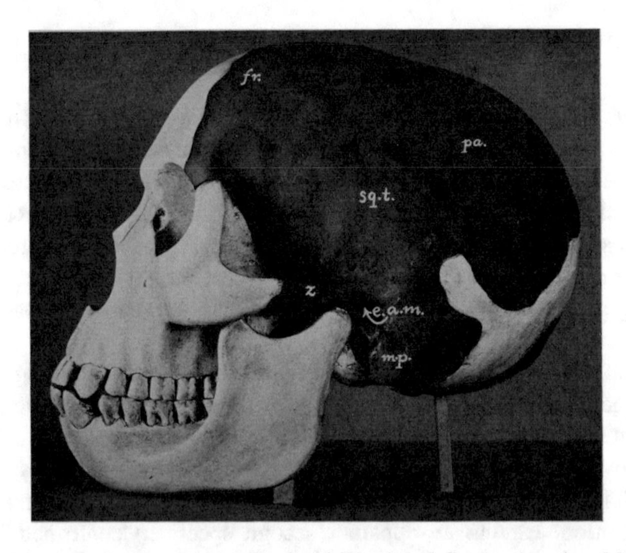

Fig. 2 The Piltdown skull as reconstructed by Smith Woodward. Source: American Museum Journal

Fig. 3 The molars on the Piltdown mandible had been filed down in imitation of heave wear. This removed characteristics that would have disguised its true identity; but on close examination, a unnatural planes of wear and striations created by a metal file exposed the hoaxer's work

nature of *Eoanthropus*. Criticism was inhibited by the ongoing World War I, which prevented foreign scientists from seeing the specimens first hand and by the untimely death of Dawson before he revealed the locality of Piltdown II.

After the war, a series of discoveries of fossil humans from continental Europe, South Africa, China, Java, and Israel began to fill the fossil gaps. Anthropologists soon realized that there was no place in the human lineage for Piltdown. Both contemporary fossils and younger ones had smaller brains and more human-like teeth. Textbooks began treating Piltdown as an anomaly and put it on a dead-end branch of the phylogenetic tree.

It was not until 1950, when Joseph Weiner reexamined the bones with a more objective eye, that the puzzle was solved. Weiner quickly realized that all the specimens from the Piltdown quarry had been painted with a stain to make them look uniformly old. The jaw and teeth had been modified to disguise the fact that they did, in fact, belong to an ape, probably an orangutan (Fig. 3). The cranium was that of a modern human with unusually thick bone. Weiner and colleagues from the British Museum convincingly argued in 1953 that the Piltdown collection was fraudulent. Further investigation showed that Dawson was almost certainly responsible, possibly with an unidentified accomplice. Woodward and his colleagues were innocent dupes.

Why Was the Forgery Accepted?

The initial acceptance of *Eoanthropus* by the scientific community raises troubling questions about the conduct and competence of science. The forgeries passed initial critical examination, and only after many years were they finally rejected. It should be noted, however, that the support for the initial interpretation of *Eoanthropus* was not as deep as common accounts suggest. A thorough reading of the literature shows that there never was consensus on how to interpret it or whether to accept it. However, many scholars and writers of prehistory did not have the opportunity, expertise, or inclination to conduct their own analyses and tended to repeat the conclusions of those who did. Given difficulty of travel and the near impossibility of comparing fossils scattered across different countries, anthropologists commonly relied on the published conclusions of the British anthropologists. Secondary and popularly written literature therefore gives a misleading impression that *Eoanthropus* was universally accepted. Nonetheless, there were many factors that made the hoax possible.

A major contribution to the success of the fraud was the particular academic circle in which it appeared. All involved, of course, were conscious of the reputation that a major new hominid skull would confer. In the context of rising nationalism that would lead to the First World War, finding an early ancestor on English soil was especially welcome to the scientists of the British Museum. Moreover, the discovery coincidentally fit neatly into previously conceived theories of human evolution. Both Smith and Arthur Keith were keenly interested in the evolution of the human brain and had argued that a large brain, being the defining human trait, must have evolved before other human characteristics. The Piltdown find appeared to confirm his prediction.

Purely from an objective perspective, however, the most persuasive argument for placing the fragments into a single skull was that the specimens were found together, though some critics saw the weakness of that argument. The forgery, though amateur in some ways, was plausible. The different bones were of a compatible size and were stained to show similar wear and coloration, yet they were incomplete enough to avoid revealing too much of their true affiliations. Misinterpretation of the remains, for example mistaking the upper canine for a lower, and the absence of established methodology for comparing species and populations contributed to the obfuscation. Moreover, the bones were accompanied by tools and parts of other animals that gave it an acceptable context.

The potential existence of Pliocene or Pleistocene remains of large-brained and modern-looking humans received qualified support from a number of other controversial discoveries of the period. Keith made a list of these to reinforce his case for Pleistocene humans that included specimens largely rejected by Keith's contemporaries as recent or of indeterminate age. Two slightly later discoveries of genuine antiquity, Steinheim (in 1933) and Swanscombe (in 1936) also combined small brain sizes and some modern features. They added enough ambiguity to the record of morphological evolution to further obscure the issue. For those who disagreed with Keith about Piltdown, there was no satisfactory alternative interpretation of the bones. Because there was no expectation of finding an ape in England in the Pleistocene, the remains must have been hominin. The discovery of Piltdown II seemed to be the final proof.

The Problems with Scientific Rigor

Although science by nature is contentious, since colleagues are expected to critique new claims and interpretations, there are reasons why the anthropologists appeared to be more accepting of the find than they were in reality. The most important of these is that science cannot reject valid data. The nature of the scientific process requires that an acceptable hypothesis take into account all the relevant evidence. When data clearly contradict it, the hypothesis must change. The only alternative is to reinterpret the data on the basis of errors or of new perspectives, or to set the data aside until better models resolve the contradiction. This principle lies at the foundation of scientific methodology, and differentiates it from the advocacy used by many nonscience disciplines.

At the time of its discovery, *Eoanthropus* was a plausible ancestor. Continued studies, however, revealed more rather than fewer inconsistencies of age, wear, and articulation between parts. The frustrations were expressed well by an American anthropologist as he conceded reluctant acceptance of the mandible:

> The prehistoric archaeologist sometimes uncovers strange bedfellows; no other discovery is quite so remarkable in this respect as the assemblage from Piltdown. Nature has set many a trap for the scientist, but here at Piltdown she outdid herself in the concatenation of pitfalls left behind – parts of a human skull; half of an apelike lower jaw, a canine tooth, also apelike; flints of a Pre-Chellean type; fossil animal remains, some referable to the Pliocene, others evidently Pleistocene; all were at least as old as the gravel bed, and some of the elements apparently were derived from a still older deposit…. (MacCurdy 1924: 333).

In spite of such misgivings, until the specimens were exposed as a forgery, they could not be dismissed. Instead, there was an extensive and imaginative effort to find an interpretation of the site that reconciled it with the steadily emerging picture of human evolution. The scientific community in this case was properly examining itself. Most researchers in the field had rejected Piltdown before 1953 by setting it aside as unresolved, so that it no longer shaped theories of human evolution; but it was only when Weiner and Oakley examined the specimens with the hypothesis of fraud in mind that Piltdown could be completely dismissed as essential fossil evidence.

Self-Correction

Formal critique began with the first description of *Eoanthropus*, as recorded in discussions at the meetings of the Geological Society. The important questions that were to be asked repeatedly over the next four decades were immediately raised: Was the reconstruction valid? Was the dating valid? Did the skull and jaw belong together? The correct answers to these questions were acquired by continued scientific study and came before 1953. No single criticism of the Piltdown specimens swayed the scientific community. Rather, the gradual erosion of the theoretical framework in which *Eoanthropus* fit led to a steady increase in skepticism between 1930 and 1950 and a decrease in the ranks of its advocates.

Much of the controversy centered on the anatomical difficulty of reconciling the human cranium with the ape-like jaw. Woodward opted for a more primitive reconstruction of the skull with a smallish brain. Keith correctly challenged Woodward's interpretation, making the brain larger and the unknown canines teeth small, like those of humans. A bitter argument ensued for many years, even after the recovery of a large canine. Other researchers converged on an anatomically correct interpretation and incidentally helped to advance the science of skeletal reconstruction. Nonetheless, the presence of a modern cranium in an ancient setting was a significant fact. With or without the mandible, in the context of later discoveries, it supported the hypothesis of multiple hominid lineages.

Relating the large cranium to the ape-like mandible was the most troubling aspect of *Eoanthropus*. Researchers attempted to resolve this anomaly in a variety of ways. If the brain were, in fact, smaller, there would be less of a problem. Smith emphasized primitive aspects of the brain while Woodward reconstructed it at the lowest end of the human range. If the mandible were more human-like, there would be fewer disharmonies. Keith initially argued for shorter jaws and more human-like dentition until the large canine was found. Other researchers lined up behind one or the other of these positions. However, the most common position of foreign scientists was to recognize two different species. As years went by, more and more anthropologists wrote the specimens off as aberrant or irresolvable (Fig. 4).

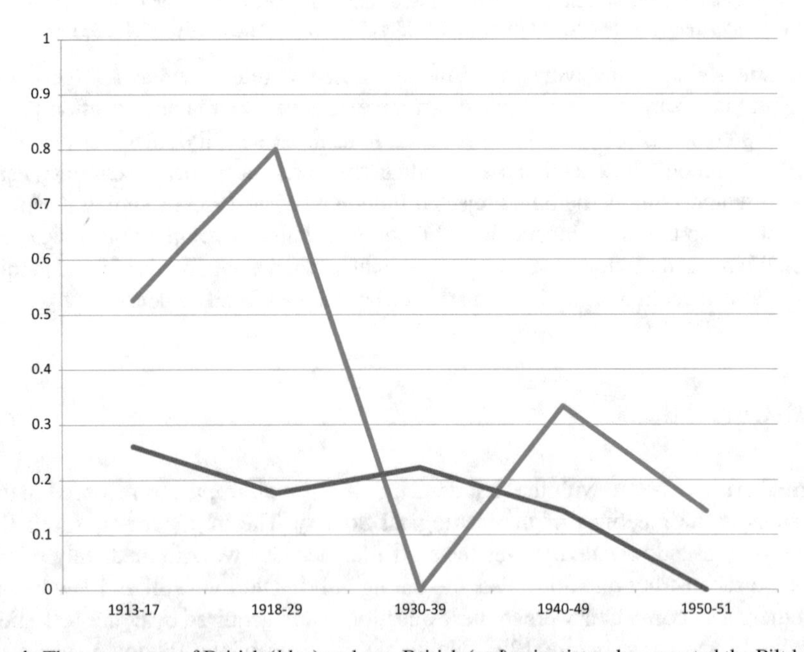

Fig. 4 The percentage of British (*blue*) and non-British (*red*) scientists who accepted the Piltdown specimens as a legitimate fossil hominin declined rapidly through time. Most foreign anthropologists expressed skepticism from the beginning. This plot is based on 110 publications between 1913 and 1953

The initial presentation by Dawson and Woodward portrayed the Piltdown individual to be a large-brained ancestor of modern humans. The primitive jaw indicated its transitional status and revealed the sequence of evolutionary change. Keith championed that view in his subsequent writings, seeing it as a confirmation of his earlier theories. Yet this interpretation faced increasing conflict from new fossil discoveries. *Australopithecus* (1924, 1936, and later), Peking Man (1921–1937), and the Kabwe skull (1921) continued to reinforce the evidence that the cranium evolved more slowly than other parts of the anatomy. The robust smaller-brained pithecanthropine lineage, into which later Neanderthals had been placed, gained importance in phylogenies, and the position of *Eoanthropus* became ever more peripheral and problematic. Discoveries at Mount Carmel between 1929 and 1934 documented an apparent transition between the Neanderthal and modern skull morphology that convinced even Keith that the large brain of Piltdown was a precocious sideshow at best.

A key issue to understanding the Piltdown specimens was their age. If they were recent, then a modern-looking cranium was not a puzzle. If the mandible and jaw were of different ages, they clearly would represent different individuals. If, however, the fossils were of similar antiquity as they appeared to be, they might represent the earliest known human ancestor. Before the 1950s, there was no technology that could reveal the absolute age of a fossil. Only relative dating techniques performed by matching the strata with others earlier or later could reveal an approximate age.

The Question of Dating

Long before the discoveries of fossils, much of Europe and America had been covered by a series of glaciers. As global temperatures declined, ice sheets advanced from the north, scouring the countryside and pushing before them walls of gravel and debris, including bones or fossils on scraped from the surface. During brief warming periods, the glaciers halted and retreated. The rows of gravel were left in place as moraines. Later, as rivers of melt water began to erode channels into the soil, gravels were washed down and redeposited at lower levels. Different erosion stages could be identified in southern England by these gravel terraces. The Piltdown site was a gravel quarry in such a secondary deposit.

There were thus several clues to the age of the Piltdown finds. The moraine provided a minimum age, since the since the fossils could have been redeposited from older soils. The other fossils found with the skull provided another check since they would probably be of the same age. Stone tool types provided another possibility to the extent that forms of tools change through time, but this was more controversial at the time of the discovery. Unfortunately, all three of these methods produced ambiguities.

Of the animal fossils, many were similar to those recently found with a primitive human jaw in Germany at Mauer: *Rhinoceros* cf. *etruscus*, *Castor fiber* (beaver),

Cervus elaphus (red deer), and *Equus* (horse). Although scientists would now consider that collection Middle Pleistocene in age, at the time it was considered Pliocene. Dawson was known to be interested in the Mauer jaw and probably was trying to simulate a site of the same age. The collection also included two younger fossils already familiar in Britain from the Pleistocene, *Mastodon* and *Hippopotamus*, so that two different time periods were present and mixed together. To whichever assemblage the skull and jaw should be assigned, the implied date was old—older than the known Neanderthals of continental Europe. The total collection appeared to rule out the possibility that *Eoanthropus* was a recent intrusive burial.

The assemblage included some stone tools, both recognizable bifacially flaked edges (i.e., flaked on both sides) and crude fragments of flint with sharp edges. The latter, sometimes called "eoliths," were already controversial. Where some prehistorians thought they saw the first glimmerings of human culture, others saw only naturally fractured rock. Not surprisingly, believers in eoliths such as Keith used them as evidence for the earlier Pliocene date, whereas skeptics argued for the more conservative Pleistocene age. Today archaeologists dismiss eoliths as naturally formed.

The age of the Piltdown fossils was of such importance—and bafflement—that the discussion continued over the next 40 years. In 1925, the stratigraphy of the area was resurveyed and it was determined that the deposits were related to a gravel terrace that was Upper Pleistocene. This was no longer consistent with the apparent age of the animals and argued for a much later redeposition.

Other dating attempts were made directly on the fossils themselves. Fossilization, the gradual replacement of organic material by minerals, takes time. It was possible even in 1913 to assess the relative amount of organic content remaining. Examination of the cranial fragments by Dawson and Woodward showed them to be heavily mineralized. It was not possible to assign an absolute age to the process, but this indicated that the cranium was indeed old. If the same test had been applied to the much younger mandible at the same time, the hoax might have been discovered right away.

Another analytical technique for chemical dating with fluorine was applied in 1950. As bones mineralize in the ground, they absorb identifiable elements, such as fluorine or uranium, from the ground water. Eventually they will reach equilibrium with their environment, containing the same amount as the surrounding rock. The principles of using fluorine content to examine age had been known since the last century, but the techniques were considered imprecise and were not commonly applied to fossils. Kenneth Oakley can be credited with reviving its application in paleontology. In 1948, he successfully demonstrated that the Galley Hill skeleton was a recent burial. Again, fluorine "dating" does not indicate absolute age of a fossil, but it can roughly indicate whether or not a bone has been buried for a long time. In this case, the test was used to determine whether the hominin bones and the other fossils from a given site were mineralized to the same degree.

Ironically, the first application of the test to the Piltdown material in 1950 showed the cranium and jaw had similar small amounts of fluorine and appeared to support the idea that the cranium and jaw belonged together. However, the Pliocene animal

Table 1 Sample results of chemical tests on the Piltdown finds (from Weiner et al. 1955). These figures illustrate the discrepancy between the cranial fragments and the mandible and teeth, as well as the mismatches among other fossils from the site

Specimen	Carbon content (%)	Fluorine content (%)
Piltdown I left frontal	7.5	0.15
Piltdown I left temporal	4.8	0.18
Piltdown II frontal	4.4	0.11
Piltdown I mandible	14.5	<0.03
Piltdown I molar	10.0	<0.04
Piltdown I canine	12.1	<0.03
Hippopotamus molar	2.2	<0.05
Cervus metatarsal	4.1	0.1
Castor molar	6.1	0.4
Elephas molar	0.1	0.8
Caprine molar	0.7	0.7

remains had considerably more fluorine and the Pleistocene fossils produced mixed results. Oakley interpreted this test as indicating that the gravel was of an indeterminate Pleistocene age and had accumulated and redeposited bones from different time periods. Although it indicated *Eoanthropus* was not Pliocene in age, the test was unable to discriminate further.

Oakley was able to repeat the test a few years later when, with Joseph Weiner, he was explicitly investigating the possibility of fraud. More sensitive methods were available to him that had smaller ranges of uncertainty. The second set of results showed a clearer difference in fluorine content between the cranial fragments on the one hand and the jaws and teeth on the other (Table 1). The cranium matched with a known Upper Pleistocene sample. Oakley suspected the cranium was older, but that the jaw and teeth were modern bone. At this time he also tested for nitrogen content. Nitrogen, which is an essential component of protein, diminishes as a bone mineralizes. This test yielded similar results: there was reduced nitrogen content in the Piltdown cranium and a Pleistocene fossil, but the Piltdown jaw and teeth and a fresh bone had higher content.

The extent of the forgery was revealed by other methods, including microscopic examination of tool marks, discovery of deliberate staining on all the bones and tools, and external evidence connecting tampered specimens with Dawson. In 1955, Weiner and his colleagues published the results of yet another chemical analysis of all the Piltdown fossils. This time they tested not only for fluorine and nitrogen, but also other minerals expected to infiltrate from the soil—gypsum, iron, chromium, and uranium. They found a wide range of readings, proving that the fossils had probably been assembled from different geographical locations. This test conclusively revealed a discordance among the bones, which was further confirmed by radiocarbon dating.

Testing the Theory of Evolution

Evolutionary biology argues that all species are connected with one another by sequences of transitional forms that may or may not be known from the fragmentary fossil record. However, when any apparent discontinuities between taxa are so easily explained away by missing fossils, is it really possible to test this fundamental prediction? Researchers can rearrange phylogenies whenever a new fossil is discovered. It is obvious to biologists that such flexibility is necessary from the nature of the task, but does it also disguise a disturbing lack of rigor in the field?

The validity of the methods of paleobiology might be better tested by the ability to recognize a true discontinuity, where a species has no relationship with earthly forms; but such a test case, according to evolutionary theory, requires an extraterrestrial fossil or an artificial one. The synthetic *Eoanthropus* qualifies as a test. The fact that the Piltdown forgery was composed from bones of real species, intended to fit into a real lineage, and placed in a plausible geological context made it that much more effective. As it happened, Piltdown was quickly recognized as anomalous and rejected by most anthropologists long before it was revealed as a fraud.

The Piltdown episode tarnished the reputations of several scientists, but may be promoted as a successful example of the scientific method in action. It illustrates the ideal of conflict and debate, ascent and dismissal of major hypotheses and minor ones, the necessity of cumulative observations, the significance of scientific consensus, and the triumph of data over incorrect theory.

Questions for Discussion

Q1: Is this case an example of successful self-correction or failure of the scientific process?

Q2: When the scientific community made mistakes, were they properly following the rules of science?

Q3: It has been stated incorrectly that the resolution of the hoax came about because of new scientific dating techniques. What is required, if not new technology, to make the kind of breakthrough that Weiner made?

Q4: How can a person forge a fossil? Should the fact that a forger is confined to working with real bones make detection more or less difficult? (For an example of a forgery not involving real bones, look up the Cardiff Giant.)

Q5. Anthropologists followed the rules of science in continuing to accept the Piltdown skull as valid, but problematic until there was good evidence that it was a forgery. How should the scientific method deal one fact that contradicts a theory? Consider how other disciplines might respond to a similar situation in these examples.

(a) A historian explains that Cortes was able to conquer Mexico because of superior technology, horses, and native superstition; but another historian points out that smallpox and other diseases were ravaging the Mexican population at the same time.

(b) A lawyer believes his client is not guilty of assaulting another man in a bar. He has three witnesses to testify that his client was a peaceful man without a temper, but a woman claims he once struck her.

(c) A psychologist believes that exposure to video games predisposes children to violence; but 34 out of 100 children who play video games more than 3 h a day have never been in a serious fight.

Q6: What lingering impact did this episode have on the study of human origins? Here are three possibilities (a) How do/should scientists respond in the future after they have been "burned" by such a mistake? (b) Did the forgery, when it was believed to be valid, lead to or reinforce ideas that remain with us after the evidence was dismissed? (c) Did the errors undermine respect for the discipline? (Creationists still refer to Piltdown to imply that much of the fossil hominin record is untrustworthy.)

Additional Reading

Dawson C, Woodward AS (1913) On the discovery of a Palaeolithic skull and mandible in a flint-bearing gravel overlying the Wealden (Hastings Beds) at Piltdown, Fletching (Sussex). Q J Geol Soc Lond 69:117–151

Keith A (1915) The antiquity of man, 1st edn. Williams & Norgate, London

Keith A (1931) New discoveries relating to the origin of man. Williams & Norgate, London

Langdon JH (1991) Misinterpreting Piltdown. Curr Anthropol 32:627–631

Langdon JH (1992) Lessons from Piltdown. Creation/Evol 31:11–27

MacCurdy GG (1924) Human origins: a manual of prehistory. Appleton, New York

Oakley KP (1950) Relative dating of the Piltdown skull. Adv Sci 6:343–344

Oakley KP, Hoskins CR (1950) New evidence on the antiquity of Piltdown Man. Nature 165:379–382

Oakley KP, Muir-Wood HP (1949) The succession of life through geological time. British Museum of Natural History, London

Spencer F (1990) Piltdown: a scientific forgery. Oxford University Press, New York

Weiner JS et al (1953) The solution to the Piltdown problem. Bull Br Mus Nat Hist (Geol) 2:139–146

Weiner JS et al (1955) Further contributions to the solution of the Piltdown problem. Bull Br Mus Nat Hist (Geol) 2:225–287

Case Study 5. Checking the Time: Geological Dating at Olduvai Gorge

Abstract The geological ages had been mapped extensively in the 1800s and early 1900s, so that paleontologists around the world could attempt to place their fossils in an increasingly global context. However, relative dating by stratigraphy only goes so far. Any attempt to assign absolute dates before the 1950s was mere guesswork and was commonly underestimated by an order of magnitude. The development of an absolute time scale required a collaboration of geologists, physicists, and paleontologists and came about in part by the inspiration of Louis S. B. Leakey. By applying radiometric dating to the volcanic layers at Olduvai Gorge and then correlating those with paleomagnetic sequences, his team established a framework that made precise dating possible for Pleistocene deposits across East Africa and beyond. Leakey's example of bringing experts of many types together on a single project has become the model for modern expeditions.

Olduvai Gorge

No discussion of human evolution can ignore the contributions of the Leakey family from Kenya. Three generations of Leakeys have been working in the field collecting fossils and tools of past hominins since the 1920s. After completing his education in England, Louis Leakey returned to his native East Africa and began fieldwork there. At a fossil site in Tanzania called Olduvai Gorge, he found the earliest known stone tools in 1931 (Fig. 1). He returned there with his family to establish a long-term field camp in 1951 and stayed for decades. For years, they found fossilized animal bones and stone tools, but no hominins. From these discoveries, his wife Mary Leakey identified the earliest known stone tool culture. Eventually, persistence paid off for Louis, and the excavations produced the first specimens of *Paranthropus boisei* and *Homo habilis*, opening new chapters in human evolution. Louis Leakey has been much criticized for his interpretations of fossil discoveries—and few of those have stood up to later evidence—but he had a genius for public relations and a vision for organizing and inspiring scientific work. Through National Geographic society publications and television specials, he made human origins and

J.H. Langdon, *The Science of Human Evolution*,
DOI 10.1007/978-3-319-41585-7_5

Fig. 1 Olduvai Gorge. Source: Creative Commons with permission

himself household subjects and inspired international funding for research. Leakey recruited three young women—Jane Goodall, Dian Fossey, and Birute Galdikas—to carry out the first long-term field studies of the great apes. He also initiated a strategy of using international teams of researchers from different disciplines to understand fossils sites. The work of his team at Olduvai finally enabled anthropologists to place human evolution onto an absolute time scale.

When Louis Leakey began his work at Olduvai, he was ungrounded in absolute time. He could and did, however, work out stratigraphy and relative age. At Olduvai Gorge, Leakey, in collaboration with geologists, mapped out the geological layers and identified faunal correlations with other sites in Africa and Eurasia. The gorge is a product of the geology of the African Rift Valley where the motion of tectonic plates has literally been tearing the continent in two for the past several millions of years. The resulting upheavals have created mountains, valleys, lakes, and volcanoes that created ideal conditions for paleontologists. Sediments containing bones collect in water channels and lakebeds to be buried by later deposits. Over time, the bones may become mineralized, turning into fossils. Continued tectonic activity raised the strata so that streams now cut into them and erosion exposes the fossils for paleontologists to find. It is not profitable to dig blindly into the ground in hopes of discovering a fossil. Instead, paleontologists spend more time walking the surface to see what has been recently uncovered. For that reason, they prefer bare desert conditions where there is little vegetation to cover the ground. The Rift Valley has great stretches of these areas in Tanzania, Kenya, and Ethiopia where many hominin fossils have been found.

For much of Olduvai Gorge, the lowest accessible stratum or "bedrock," was a lava flow. A lake formed on top of that, collecting tens of meters of sediment. Bones and tools also accumulated at the edge of the lake. In later times, tectonic forces elevated the lakebed. As the surface rose, a river flowing across it cut through the rock and created the present day canyon. There, the Leakey family and their workforce continually surveyed the walls for fossils eroding from the rock.

Leakey identified four major beds. Bed I, at the bottom of the canyon, lies above and below the lava flow and is divided by it into Upper and Lower Members. It was in the Upper Member, in 1959, that his wife Mary Leakey discovered a cranium of a robust australopithecine that received the name *Zinjanthropus boisei* (later to be called *Paranthropus boisei*). A few years later, Leakey announced another new hominin, *Homo habilis*, also from Bed I. Scattered on various levels within this layer of rock were numerous stone tools, more primitive than any that had been found in Europe or anywhere else. The Leakeys named this earliest tool tradition the Oldowan Culture, from a variant spelling of the site. The three higher beds contained even more assemblages of tools, assigned to the Oldowan or Acheulean traditions. Bed II was later to reveal bones of both *P. boisei* and *H. habilis*, as well as *H. erectus*.

At Olduvai, Leakey claimed to have found the very origin of humanity, as signified by these earliest tools, the remains of *H. habilis*, and the bones of its prey. Their age became an important question. Leakey knew that the lava deposit at the bottom of the exposed layers probably dated to the early Pleistocene. In a 1954 article, he suggested the beds spanned a period from 400,000 to 15,000 years ago. By that time, however, more sophisticated techniques for dating were being developed that could be readily applied at Olduvai.

Radiometric Dating

In the first half of the twentieth century, physicists in Europe had learned about radioactivity and the predictable rate of decay of unstable atoms. By the 1950s, they were beginning to apply this understanding to dating minerals by the products of radioactive decay. Calculating the date of an object requires that the substance had clear beginning and that it changed in some way at a constant and known rate. Igneous rock has those properties.

When a volcano erupts, it brings forth a variety of minerals from deep within the earth. Gases are released and new rocks are formed. One element commonly present is potassium, including its radioactive isotope ^{40}K. This isotope decays into argon (^{40}Ar). Like all radioactive decay, this happens at a known, constant rate, proportionate to the amount of potassium present. Half of the ^{40}K atoms convert to ^{40}Ar every 1.2 billion years. This is the half-life of ^{40}K. Argon is a volatile gas and is lost during the eruption, but it becomes trapped in the cooled layer of ash; thus, any ^{40}Ar present in a volcanic rock has accumulated since the original eruption. Because we can measure the amount of both ^{40}K and ^{40}Ar in the rock, the ratio allows us to calculate the time that has elapsed since the eruption.

This is the basic principle behind all radiometric dating. Unstable isotopes break down at rates that are different for different isotopes but constant in proportion to the number of atoms present. If we know the original amount of one side of the equation (in this case, zero ^{40}Ar) and can measure the other quantities, it is possible to date the object in which they are contained. Uranium-containing rocks can also be dated by the breakdown of unstable isotopes. Uranium-234 decays to Thorium-230 and has a half-life of 245,000 years. This is one commonly used event in a longer series by which the uranium eventually becomes lead. Any of the decay events potentially can be used for dating uranium-bearing rocks. The better-known process of carbon-14 dating relies on the fact that carbon, including the stable ^{12}C and ^{13}C isotopes as well as the unstable ^{14}C, is taken from the atmosphere by living plants and then transferred to the animals that eat them. After an animal or plant dies, ^{14}C decays to ^{14}N, enabling us to calculate the date of the organism's death by the disappearance of ^{14}C.

There are limitations to radiometric dating. Samples that are too old may not have enough of the original isotope present for accurate measurement. Samples that are too young may not have enough of the daughter isotope accumulated. Not all materials we would like to date contain useful radioactive isotopes. Nonetheless, we can date many types of rock formations and even establish the age of the earth itself.

When Leakey learned about the newly developed K–Ar dating technique, he invited two geologists, Jack Evernden and Garnett Curtiss, to apply the technique at Olduvai Gorge. The geological conditions were right for accurate dating. The frequent volcanic eruptions in the area had deposited many blankets of ash. These ash layers, called tuffs, were datable by the K–Ar method. No less than six tuffs lie in Bed I. Evernden and Curtiss took samples, analyzed the argon content, and published them in 1961. The second lowest tuff, called Tuff 1B, lay immediately under the *Zinjanthropus* skull. When it was dated, it proved to be 1.75 Ma, which was astounding at the time. This date put an absolute time frame onto human ancestry and helped calibrate the start of the Pleistocene at just under 2 Ma ago.

Paleomagnetism

Radiometric dating was not the only dating technique developed at Olduvai. In the 1950s, while some geologists were experimenting with radiometric dating, others were exploring paleomagnetism. As certain types of sedimentary rocks form, iron-containing particles align themselves with the earth's magnetic field. This discovery was accompanied by observations of rocks that were out of alignment. Either the rocks or the magnetic poles had moved subsequent to the formation of the rocks. In reality, both happened, as appreciation of continental drift made clear. More intriguing was the discovery that the poles occasionally reversed.

On the Atlantic sea floor, new rocks form along a central rift and subsequently are pushed to the east or west. When, periodically, north and south magnetic poles reverse, new particles being deposited change their alignment accordingly. On the sea floor where rocks are created, we can observe a record of the earth's magnetic history spread laterally on either side of the rift (Fig. 2). Where deposition in one

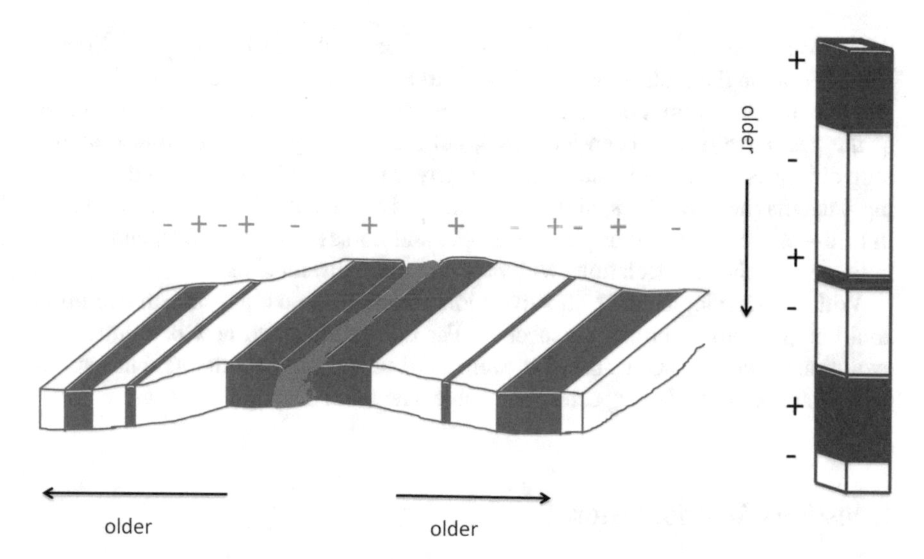

Fig. 2 The earth's magnetic orientation is recorded in sediments as they form. In the mid-Atlantic Rift, an upwelling of magna is creating new crust on both the east and west, pushing the continental plates away (*left*). Past magnetic normal (+) and reversed (−) orientations may be observed as one traverses older rocks. In other areas, such as Olduvai Gorge, continuous deposition creates a vertical sequence (*right*)

place has been continuous over long periods of time, those rocks show us a detailed record in a vertical column. Such sediments are present on the ocean floor because of the constant rain of particles from above. Deep sea cores enable us to reconstruct the history of paleomagnetic reversals in fine detail. We have learned that the magnetic poles, for reasons not fully understood, flip at irregular intervals of tens of thousands to tens of millions of years.

Most terrestrial layers of rock are not continuous, but have many gaps representing periods of time when the surfaced eroded. It is difficult to know how long a time may be represented by a discontinuity or what paleomagnetic events might have been lost in that interval. One important exception is Olduvai Gorge, where deposition was reasonably continuous through most of the last 2 Ma. Consequently Olduvai was a good place to begin mapping the magnetic orientation in terrestrial deposits.

Although the terrestrial sequence should exactly follow that seen in deep sea cores, differences in sedimentation rates and quality of samples and improving technology make each new study a valuable addition to our knowledge. When information about paleomagnetism is added to the radiometric dates, magnetic reversals become datable historic events. At Olduvai, geologists discovered a reversal event. Named the Olduvai Event, it is a brief period of normal polarity lasting about 150,000 years in the midst of a longer reversed period called the Matuyama Epoch. The Olduvai Event lies at the end of the Pliocene epoch, just after 2.0 Ma. When sediment conditions are complete, the Olduvai Event is an important chronological marker anywhere on the earth.

In the following decades, the Leakey family and other anthropologists, especially from France and the United States began to explore other fossil localities throughout East Africa, including the Omo River valley in Ethiopia and Kenya; Koobi Fora, on the eastern shore of Lake Turkana; and the Afar region of Ethiopia. At Omo, dozens of tuffs had been created over a 3 Ma period. The great volcanoes of the past have not all been identified, but their ash layers can be traced across long distances. This fine volcanic stratigraphy, dated by radioisotopes and interlacing paleomagnetic markers, allows the dating of fossils of all kinds. Those fossils, in turn—especially those of pigs, antelope, and horses—became independent age indicators to check correlations with new and more distant sites.

With the development of effective tools for absolute dating, human evolution could be put into its proper perspective. Far more time could now be allotted for evolutionary change. Our ancestors were not hurrying to become us, but experienced long periods of unexpected, but stable adaptations along the way.

Questions for Discussion

Q1: Why can't we assign reasonable dates to all fossils?

Q2: Many important fossil specimens are extremely hard to date, because their context was not datable, they were found not in situ, or they were recovered before modern technologies were established. How does this limit their value in understanding human evolution?

Q3: What is the difference between a margin of error in calculating an age and an unreliable date? Does one imply the other?

Q4: What is the difference between precision and accuracy in dating?

Q5: There are many advantages in collaborating with colleagues in different disciplines. What are the realistic barriers to collaboration?

Q6: As interdisciplinary teams of scientists becomes a more common way of operating, is it still necessary for individuals to be trained in many fields instead of specializing in one?

Q7: There are many archeological mysteries that might be addressed if we know the dates of objects. Consider the following controversial objects/events: the Shroud of Turin; the origin of Stonehenge; the sarcophagus of James, brother of Jesus; the first arrival of people in the New World. Why are these so difficult to date?

Additional Reading

Hay RL (1990) Olduvai Gorge: A case history in the interpretation of hominid paleoenvironments in East Africa. In: Laporte LF (ed) Establishment of a geological framework for paleoanthropology. Geological Society of America Special Paper 242, 23–37

Johanson D, Shreeve J (1989) Lucy's child. Avon Books, New York

Leakey LSB (1954) Olduvai Gorge. Sci Am 190(1):66–72

Leakey MD (1971) Olduvai Gorge, vol 3, Excavations in Beds I and II, 1960–1963. Cambridge University Press, Cambridge

Leakey LSB (1976) By the evidence: memoirs 1932–1951. Harcourt, New York

Case Study 6. Quantifying Evolution: Morris Goodman and Molecular Phylogeny

Abstract The classification of living animals has long relied on identifying similarities and differences in anatomical traits in adults and on developmental stages in embryos. It is assumed that such traits reflect the genes of the individual and thus its lineage. In the twentieth century, it became possible to compare species on the basis of molecules. This provided a new and independent means to test the conclusions based on morphology. The results reaffirmed our general understanding of taxonomy, but overturned our understanding of the place of one species in particular—*Homo sapiens*. Molecular anthropology has now become an essential dimension of any attempt to understand human evolution. This chapter and the next look at the immediate impact of molecular studies in understanding the relationship between humans and living apes. We consider this revolution first from the molecular viewpoint and in the next case study from the fossil perspective.

The classification of animals and plants has been a significant obsession of naturalists since classical times. They recognized intuitively that some organisms belong together—those with green leaves, those with hair and teeth, or those who kill and eat other animals. However, these self-evident groupings had no theoretical basis until the coming of evolutionary theory. In the Darwinian paradigm, each taxon—species, genus, family, or kingdom—is supposed to represent a group of organisms descended uniquely from a common ancestor.

In the past century, we have been more deliberate and systematic about this. When a given species acquires a new character trait, it is likely to pass that trait on to its descendants. We can therefore recognize the relatedness of two organisms by the fact that they have traits in common that were not present in a more distant ancestor. Scientists call these shared derived characters. For example, we know mammals are related because they have mammary glands that were not present in the synapsid reptiles from which they are descended. The first mammal species to evolve this character passed it on to its descendants. The absence of mammary glands would be considered a primitive state in this lineage, and the presence of them is derived. Mammals today also have many other traits in common—hair and complex teeth, for example—that were acquired by the early common ancestor. Scientists use these shared derived traits to determine whether an animal is or is not a mammal.

J.H. Langdon, *The Science of Human Evolution*,
DOI 10.1007/978-3-319-41585-7_6

With this simple rule, it would seem very easy to produce an accurate classification. Unfortunately it is not that easy. Sometimes derived characters are lost, as hair is lost among the whales and dolphins. Sometimes derived characters evolve independently in parallel, within different lineages. For example, both flying squirrels and flying lemurs have skin flaps attached to their limbs that enable them to glide considerable distances from tree to tree, yet they evolved these independently and therefore skinfolds do not indicate close relationship. In fact, it is often very difficult to determine which states are primitive and which are derived, or which are shared and which are not. It is even difficult to define an independent character trait, since many are related by being under the influence of the same genes or developmental pathways. As a consequence, the fine details of taxonomic classification are areas of continuous argument and ambiguity. Biologists often argue for one classification over another by amassing greater numbers of traits that appear to have developed independently. Therefore, in addition to using anatomical features, biologists have also drawn upon embryological development, physiology, geographical distribution, ecology, and behavior to help define taxa. In the last century, a new source of data has become available—molecular structure.

Applying Molecules to Classification

From the early 1900s, physical anthropologists used crude techniques such as blood typing and electrophoresis to study molecular variation among human populations. Blood types were discovered about the turn of the century from the immune responses they could evoke in individuals of other blood types. Electrophoresis was used to separate molecules in a gel and provided a crude, yet simple technique to sort proteins by size and electrical charge. These tools identified differences among individuals, but had not yet proved useful for classifying populations.

In parallel with these studies on humans, a few anthropologists explored differences in serum proteins among primates. Early attempts to examine relationships among species were limited by inconsistent laboratory methods and standards. Consequently, they tended to produce contradictory and confusing results. The first reliable and systematic assessment of quantitative differences was conducted by Morris Goodman in the early 1960s.

Goodman used the mammalian immune system to assess similarities of molecules in different species. When our systems encounter foreign proteins, such as molecules on the surface of a bacterium, we begin to manufacture antibodies against them. An antibody is a protein whose configuration allows it to attach firmly to the foreign protein, or antigen, and target it for destruction by immune cells in the body. If that antigen is still attached to a bacterium, for example, the body will destroy the germ as well. The fit between an antibody and antigen is precise and specific. The antibody will bind only with that antigen or with a molecule with nearly the same shape.

If a small amount of human protein is injected into a healthy laboratory animal, such as a rabbit, the rabbit will suffer no ill consequences, but its immune system will begin to manufacture antibodies against the human antigen. Subsequently, blood drawn from the rabbit will contain the antibodies. By combining the rabbit blood with additional human protein, a researcher can observe the reaction between the two under a microscope. The corresponding protein in a chimpanzee is very similar, but not identical to that of a human. It will also bind with the rabbit's antibodies, but the reaction will be less intense because the slightly different shape of the chimpanzee protein does not permit the antibody to bind as closely.

Goodman used the common blood protein albumin to stimulate antibody production in chickens and rabbits. He then compared the degree of reactivity between antibodies made for human albumin and the albumin from a number of living primates. On a triangular gel, he permitted proteins from two primate species and antibodies from the serum of a sensitized animal to diffuse into one another. The more intense reaction visually dominated the plate and created a "tail," a streak of bound antibody across the plate. Goodman described the relative intensities of an immune reaction qualitatively according to the length of a tail—trace, short, medium, or long. With a series of such qualitative comparisons, he was able to construct a phylogenetic tree of primates based on the similarity of proteins. Chimpanzees and gorillas showed the strongest reactions and thus are most closely related to humans. Orangutans are more distant. Gibbons and siamangs, closely related "lesser apes," are placed further away and six species of Old World monkeys are the least closely related to humans. This arrangement was exactly what was expected, except for the position of our species. The latter surprise was to have significant consequences for our understanding of human evolution.

Up to this time, the three great apes—chimpanzees, gorillas, and orangutans—were classified in Family Pongidae. They obviously shared a number of traits to justify this, including hairiness, semi-upright body posture, long arms and short legs, a relatively long face, and large canines. Humans were placed alone in a family of our own, Hominidae. Certainly our suite of unique characters, including hairlessness, bipedalism, and large brains justified such a distinction, if not something higher, such as a separate phylum or even kingdom, as had been suggested. In a narrower evolutionary framework, this classification implicitly stated that the great apes shared a common ancestor more recently than the ape–human divergence. Goodman's results, on the other hand, indicated that the orangutan diverged first and that humans, chimps, and gorillas descended from a later common ancestor. They argued for a new scheme of classification. Molecular anthropology was set up for a collision with traditional wisdom.

Shortly thereafter, Vincent Sarich and Allan Wilson repeated his studies and introduced a quantitative measure they called an "immunological distance" or "index of dissimilarity" between two species. The index was the ratio of quantities of antiserum needed to create the same intensity of reaction to proteins of two different species. Their data confirmed Goodman's observations on the relationships among higher primates (Table 1).

Table 1 Indices of dissimilarity for hominoid albumins (Sarich and Wilson 1967). A value of 1.00 indicates the proteins of two species are effectively identical

	Antiserum to *Homo*	Antiserum to *Pan*	Antiserum to *Hylobates*
Homo, human	1.00	1.09	1.29
Pan troglodytes, chimpanzee	1.14	1.00	1.40
Pan paniscus, bonobo	1.14	1.00	1.40
Gorilla, gorilla	1.09	1.17	1.31
Pongo, orangutan	1.22	1.24	1.29
Symphalangus, siamang	1.30	1.25	1.07
Hylobates lar, gibbon	1.28	1.25	1.00
Old World monkeys (average of six species)	2.46	2.22	2.29

Goodman produced important results using technologies that would be considered very crude by modern standards. Like his paleontologist counterparts, he was trying to comprehend genealogies on the basis of very indirect evidence. Comparative anatomists look at the form of the body as a proxy for the genetic coding that lies behind it. The molecular biologists were interested in the forms of proteins for the same reason. Behind the concept of a genetic phylogeny is the proposition that simply counting the differences in the accumulated number of mutations among multiple species will enable us to map those species onto a phylogenetic tree. Yet the shapes of bones and the shapes of proteins are only indirect reflections of the DNA sequence. What was needed was a more direct way of examining and comparing genes. Subsequent decades have provided that technology, culminating in the ability to sequence long strands of DNA. The Human Genome Project has mapped large parts of human chromosomes and many individual genes are known in detail. In order to make use of this information for understanding evolution, we must be able to compare the data with that from other species. The number of species whose genome has been mapped at least on some level is increasing rapidly.

Molecular data is not, of course, an alternative approach to classification, but a complimentary one. Each nucleotide may be considered an independent character trait, but not very reliable by itself. A single nucleotide on a chromosome has only four possible character states, depending on which of the four bases of DNA occupies that site—either guanine, cytosine, adenine, or thymine. We assume that if the corresponding, or homologous, site is occupied by the same base, this is a shared derived character from the last common ancestor. That may be the case, but if there have been multiple mutations at that locus in the past, it is possible that the similarity is coincidental. With a limited number of discreet character states, it is relatively easy for multiple mutations to return the nucleotide to its original form, thus erasing part of its evolutionary history. While anatomists found that looking at many traits produced more reliable results than looking at a few, geneticists must look at the most probable interpretations from the analysis of thousands or millions of independent data points from long stretches of DNA.

A New Classification

The most closely related species should be placed in the same family. If we are to recognize more than one family for the living great apes and humans, the outlier is genus *Pongo*, the orangutan, which can remain in its own Family Pongidae. *Gorilla* and *Pan* should be transferred to Family Hominidae, with us (Fig. 1). While this concept has been fully embraced by the anthropological community, we still have some trouble adapting the terminology. For a century, we have used the common term hominid for members of Family Hominidae, referring only to ourselves, *Homo sapiens*, and our fossil relatives, and pongids for the great apes. Now, "hominid" also includes gorillas and chimpanzees. There is no formal taxonomic term that encompasses the three living genera of great apes. Humans and fossil relatives are properly placed in Subfamily Homininae and called hominins. It is obviously impossible to rewrite a century of literature, and until the 1990s many publications continued to use the more familiar term hominid.

The last issue to be resolved was to sort out the exact relationships among humans, chimpanzees, and gorillas. Early studies found that, within the limits of

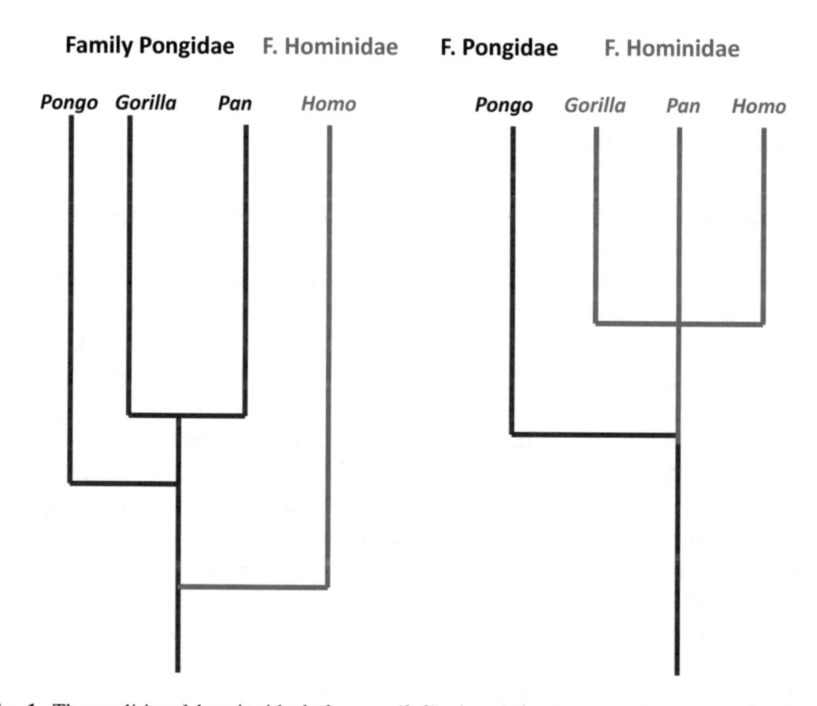

Fig. 1 The traditional hominoid phylogeny (*left*) placed the three great apes together in the family Pongidae and humans in our own family. The molecular phylogeny worked out by Goodman (*right*) found that humans were more closely related to the African apes and should be placed in the same family with them

resolution, all three lines were equally related. Data could be found to support any pairing, and the possibility was entertained that the three lineages did indeed split simultaneously. Eventually, the examination of more and different gene sequences showed that gorillas diverged first and that chimpanzees are humans' closest relatives. More surprising is the conclusion that humans, not gorillas, are chimpanzees' closest relatives. However, the differences are small. A nucleotide or a single gene or any sequence of DNA tells only the history of that gene or sequence, not of the species. Therefore, it is not surprising that studies gave different results. Moreover, speciation, the creation of a new species (in this case by dividing one population into two) takes a substantial period of time. The two speciation processes producing three African lineages—humans, chimpanzees, and gorillas—may have overlapped in time. Resolving them into earlier and later events may be artificial and misleading.

The philosophical implications of this change have been as substantial as the biological and terminological ones. Darwin's work undermined the belief that humans were unique by locating us in relationship to all other life forms. Molecular studies took this a step farther. No longer could anthropologists think in terms of a long separate human lineage, perhaps going back to the Oligocene. Instead, humans were placed among the great apes, not outside of them. Humanity's evolutionary divergence occurred over a much shorter period of time, and presumably to a lesser extent, than had been imagined.

Questions for Discussion

Q1: Why would classification of animals have been so difficult in the centuries before Darwin?

Q2: The scientific discovery and exploration of the non-European world, which was intensely pursued in the eighteenth and nineteenth centuries, resulted in the discovery of thousands of new plant and animals previously unknown in Europe. What effect might this have had on attempts to sort organisms into meaningful taxa?

Q3: There has been a tension between traditional, more intuitive classification systems that attempt to recognize degrees of distance (such as the distinctiveness of humans) and a more formal system called cladistics, which only recognizes direct genealogical relationships. The events in this chapter occurred as cladistics was replacing traditional classifications. What are the advantages and disadvantages of each?

Q4: Why is a molecular approach likely to be more objective than an anatomical one?

Q5: When a new methodology produces results that contradict those of an established technique, we should approach it carefully. Under what circumstances should the new method be accepted and the old ideas revised?

Additional Reading

Goodman M (1963) Serological analysis of the systematics of recent hominoids. Hum Biol 35:377–424

Goodman M (1967) Deciphering primate phylogeny from macromolecular specifications. Am J Phys Anthropol 26:255–275

Marks J (1994) Blood will tell (won't it?): a century of molecular discourse in anthropological systematics. Am J Phys Anthropol 94:59–79

Marks J (1996) The legacy of serological studies in American physical anthropology. Hist Phil Life Sci 18:345–362

Sarich VM, Wilson AC (1967) Immunological time scale for hominid evolution. Science 158:1200–1203

Case Study 7. Reinterpreting *Ramapithecus*: Reconciling Fossils and Molecules

Abstract In 1967, two important papers were published that had bearing on the start of the hominin lineage. One sorted and reclassified the fossil record, promoting *Ramapithecus punjabicus* as the earliest known hominin. The other used molecular comparisons of living primates to calculate the time the hominin lineage diverged from other hominoids. These two studies incompatibly disagreed over the timing of that split, but at the time both conclusions represented the best interpretations of different lines of evidence. The struggle to reconcile them stimulated new research and profoundly changed the way we understand ourselves.

The classification of living animals has long relied primarily on identifying similarities and differences in anatomical traits in adults. A trait uniquely shared among a group of animals may represent common descent from the first animal to display that trait—for example, if monkeys, apes, and humans all have color vision, then the common ancestor of monkeys, apes, and humans also had color vision. The alternative to such a hypothesis is the assumption that the trait evolved independently in the different lineages. In the absence of other evidence, we would consider the first explanation to be more parsimonious, because it requires only one evolutionary event instead of many. It is assumed that anatomical traits reflect the genes of the individual and thus its lineage, but there are many possibilities for misinterpretation that can create ambiguity. Current practice thus attempts to classify organisms on the basis of many such shared derived traits.

Traditional views from the 1800s placed the great apes together in a single taxon, Family Pongidae. Orangutans, gorillas, and chimpanzees have many similarities. They are large hairy arboreal primates exhibiting some degree of prognathism (facial elongation) and long sexually dimorphic canines, but these are primitive characters and should not be used for classification. However, the great apes have many derived characters in common: they have expanded brains and enhanced intelligence. The ribcage is flattened from front to back and the shoulders are oriented laterally to allow the animals to reach to the side of overhead. Their upper limbs are proportionately long and are used to climb and hang from tree branches. The lower limbs are short, but are frequently used to support the body in upright positions for

J.H. Langdon, *The Science of Human Evolution*,
DOI 10.1007/978-3-319-41585-7_7

standing bipedally or climbing. Both hands and feet have long grasping digits and mobile joints to facilitate climbing. In all of these derived characters they resemble humans; but people are so different from the apes that they were placed in a family of their own, Hominidae. The evolutionary meaning of that classification, enthusiastically endorsed by interpretations of the very scanty fossil record, was that the human lineage diverged from the great ape line in the distant past, well before the three great apes themselves became distinct.

The Molecular Clock

Vincent Sarich and Allan Wilson took Goodman's technique for classifying species by molecules (Case Study 6) a step further by quantifying the observations of immunological distances. They also figured out a way to calculate the time in the past at which each of these lineage splits occurred and pioneered what is known as the molecular clock. While its ability to tell time with great precision is a matter of continuing debate and research, the clock has changed the way we investigate evolution.

The essence of any clock or dating technique is a pacemaker, some element that changes at a constant and known rate. In a grandfather clock, it is the swing of the pendulum by the unchanging force of gravity, in an electric clock, the alternation of current 60 times per second. Sarich and Wilson proposed that mutations accumulate at a constant rate—constant, at least, when averaged over millions of years of evolutionary time. If a splitting event that was well documented in the fossil record could be compared to the immune distance between the two lineages, it would be possible to calculate the rate of molecular change, and thus the divergence dates for all these lineages.

The calibration point for the clock was the divergence of apes from Old World monkeys. This was thought to have occurred about 30 Ma ago, based on the fossils from the Fayum site in Egypt that were believed to be the earliest representative of these groups. The immunological distance that had accumulated in both lineages during the past 30 Ma allowed Sarich and Wilson to calculate a rate of change. Using that rate, they concluded that the differences among humans, chimpanzees, and gorillas that had accumulated since the last common ancestor would only have taken about 5 Ma.

Understanding why the molecular clock works requires some understanding of how genes change. Why should mutations be expected to occur at a constant rate? One model attempting to answer that question proposes that the vast majority of evolutionary change is selectively neutral. That is, the mutations neither increase nor decrease fitness. Once they occur, simple chance allows them to become more common or simply to disappear. The random sampling of parental genes that occurs with the conception of each new generation is called genetic drift, and it may account for most of the loss or fixation of variations from the population. Since mutations themselves occur unpredictably, but overall at a fairly constant rate, the rate at which new mutations become fixed is likewise roughly constant. This rate, in turn, drives the molecular clock.

In the neutralist model, natural selection is an important but relatively rare agent of change. A competing model views selection as the primary mechanism of change, with selection being constant and unrelenting. Since the environment—including predators, parasites, prey, conspecifics, physical conditions, and other genes in the same organism—changes constantly, species and their genes must also change continually or perish. This model is called the Red Queen Hypothesis after the character in *Through the Looking Glass* who says to Alice, "... you see, it takes all the running you can do, to keep in the same place."

Nonetheless, there are valid theoretical reasons to expect the rate of molecular change to vary. For example, chromosomes are well protected from mutations most of the time, except when they are replicating. Since only mutations in the sex cells or the germ line are relevant here (others are not passed to future generations), the opportunity to introduce new mutations occurs primarily only when that cell line is being created during early development or when the animal is reproductively active and sperm are being produced. In a given length of time, there will thus be more opportunities for mutation in a species with a short generation length than in one such as humans that matures slowly. We now know that rates of change can vary in different lineages and that variation is partly explained by generation time. Other factors may also be at play, as well, including body size and metabolic rate, and it is fairly common for the molecular clock to disagree with the fossil record.

Apes of the Miocene

Before 1967, the taxonomy of the fossil apes had been in chaos, with no less than 28 genera and 53 species of medium and large-bodied hominoids named from the Oligocene and Miocene, as well as a number of smaller species. It had been routine for new specimens to be given new names, regardless of very fragmentary condition. Fossils from Africa, Europe, and Asia were widely scattered in collections around the world, and direct comparisons were impractical. The person to sort out the redundant classifications was paleontologist Elwyn Simons, who had restarted excavations at the Fayum. With his student David Pilbeam, Simons simplified the classification of Miocene and later hominoids to three genera and ten species. He believed these had all descended from an Oligocene species he had discovered at the Fayum, which he named *Aegyptopithecus zeuxis*. Moreover, Simons and Pilbeam paired orangutans, gorillas, chimpanzees, and humans with specific Miocene species (Fig. 1). The African apes, they thought, were apparently related to African species of *Dryopithecus* about 22 million years old, whereas orangutans and humans could be linked with two species from Asia, *Dryopithecus sivalensis* and *Ramapithecus punjabicus*, both about 14 million years old. After that date the fossil record of hominoids was a blank until acknowledged hominins, of genus *Australopithecus*, showed up 11 Ma later in the record.

Simons declared *Ramapithecus* to be the human ancestor from the Middle Miocene. It had been discovered and named from jaw fragments found in the Siwalik Mountains of northern India in 1911 by Guy Ellock Pilgrim. By the time

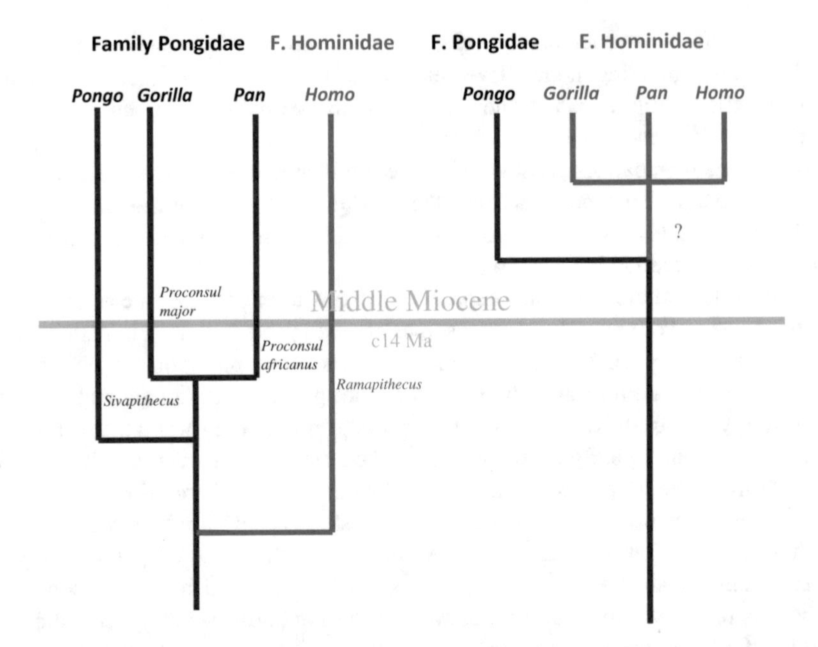

Fig. 1 A hominoid phylogeny based on fossils (*left*) identified an early separation of the human lineage and all living genera distinct by the Middle Miocene. A phylogeny based on the molecular clock (*right*) identifies a late split for humans and a close relationship with chimpanzees and gorillas

Simons reexamined the material, a few more partial jaws were known, each under a different species or genus name. *Ramapithecus* could be associated with later hominins on the basis of several characteristics, including relatively large molar teeth and a robustly built mandible. More specifically, he identified three key characters uniquely shared with humans: thick enamel, reduced canines, and a parabolic dental arcade. Humans and our near fossil relatives share a relatively thick layer of enamel on our teeth compared with monkeys and living apes. This enables our teeth to withstand greater bite forces and to last longer. The presence of long knife-like canine teeth, especially in males, is a general primate trait, but is reduced in humans and known fossil relatives. The shape of the dental arcade, or tooth row, is also distinctive. In other primates, the rows of cheek teeth run parallel to one another. The large canines at the front form corners of a U-shaped or box-shaped arcade. Rows of cheek teeth in humans diverge from front to back to form a parabola. Enamel thickness, canine length, and arcade shape were distinctive derived characters unique to humans that could identify our fossil relative. *Ramapithecus*, known from several small fragments of jaw, was such a relative, if we can assume the characters evolved only once. This interpretation was not without its critics. Leonard Greenfield, among others, argued that *Ramapithecus* resembled *D. sivalensis* closely, differing primarily in ways that differentiate male apes from females. Nonetheless, the Simons and Pilbeam model was widely accepted and appeared in textbooks.

Also in 1967, as Simons and Pilbeam simplified the Miocene taxonomy, Vincent Sarich and Allan Wilson published their molecular clock. The molecular date of

5 Ma for the human–ape split was incompatible with the interpretation of the fossils. A 14-million-year-old fossil in Asia could not possibly be a human ancestor unless it was also an ape ancestor. Sarich wrote bluntly in 1971 words certain to raise the hackles of the paleontologists: "One no longer has the option of considering a fossil older than about 8 Ma as a hominin no matter what it looks like." Anthropologists responded to these competing models vigorously. Both sides had made falsifiable predictions about the fossil record. It was clear that the best way to resolve the issue was to find new fossils. With more funding, better access to sites around the world, better questions, and more anthropologists, the pace of fieldwork in Africa, Europe, and Asia accelerated. Eight new genera of hominoids were named in the 1970s. The numbers would increase by more than a dozen in each of the following decades. Today, about 50 genera of fossil hominoids and early anthropoids are recognized, but anthropologists continue to disagree on details of taxonomy and on which specimens are recognized as distinctive species or genera. As the new specimens and new taxa were placed into the existing framework, it was quickly realized that there were far more Miocene apes than anyone had expected.

New Discoveries from the Siwalik Mountains

The discoveries that would help resolve the controversy came from Pakistan. *Ramapithecus* had been discovered in the Siwalik Mountains of India, and it was natural to return to this region to search for more evidence of the origins of the human lineage. David Pilbeam began leading expeditions to an extension of these mountains in neighboring Pakistan in 1973. The extensive mammalian fossils from the Siwalik area showed a forested habitat. Hominoid fossils turned up regularly—nearly a hundred specimens had been found in India; however, these were small fragments and difficult to sort into species. A steady trickle was added over the following years. A better understanding of the geology led to a new estimate of the date for this material of about 9 Ma, later adjusted to eight. This lessened the discrepancy between the fossil and molecular clocks, but still left a sizable gap. However, the fossils began to tell a different story.

As more Asian *Dryopithecus* material accumulated, it was apparent that it resembled *Ramapithecus* in its robustness and in the thickness of the enamel. It differed from *Dryopithecus* species from Europe and Africa in these same characteristics, so the older name *Sivapithecus* was resurrected for the Asian specimens. Moreover, new material of *Ramapithecus* looked more and more like *Sivapithecus* (Fig. 2). GSP 4857 is a relatively complete mandible missing most of its teeth that was referred to *Ramapithecus* on the basis of size. The rows of tooth sockets on each side diverged markedly but are straight, not parabolic. GSP 9977 is a nearly complete palate with a large canine. Its tooth rows are straight and parallel to one another. The previously observed distinction between U-shaped and parabolic-shaped arcades was complicated and ultimately discarded as not very meaningful. The apparent rounding of the human jaw is largely a secondary effect of smaller canines and shorter jaw. GSP 9564, a mandible assigned to *Ramapithecus,* had large sockets for

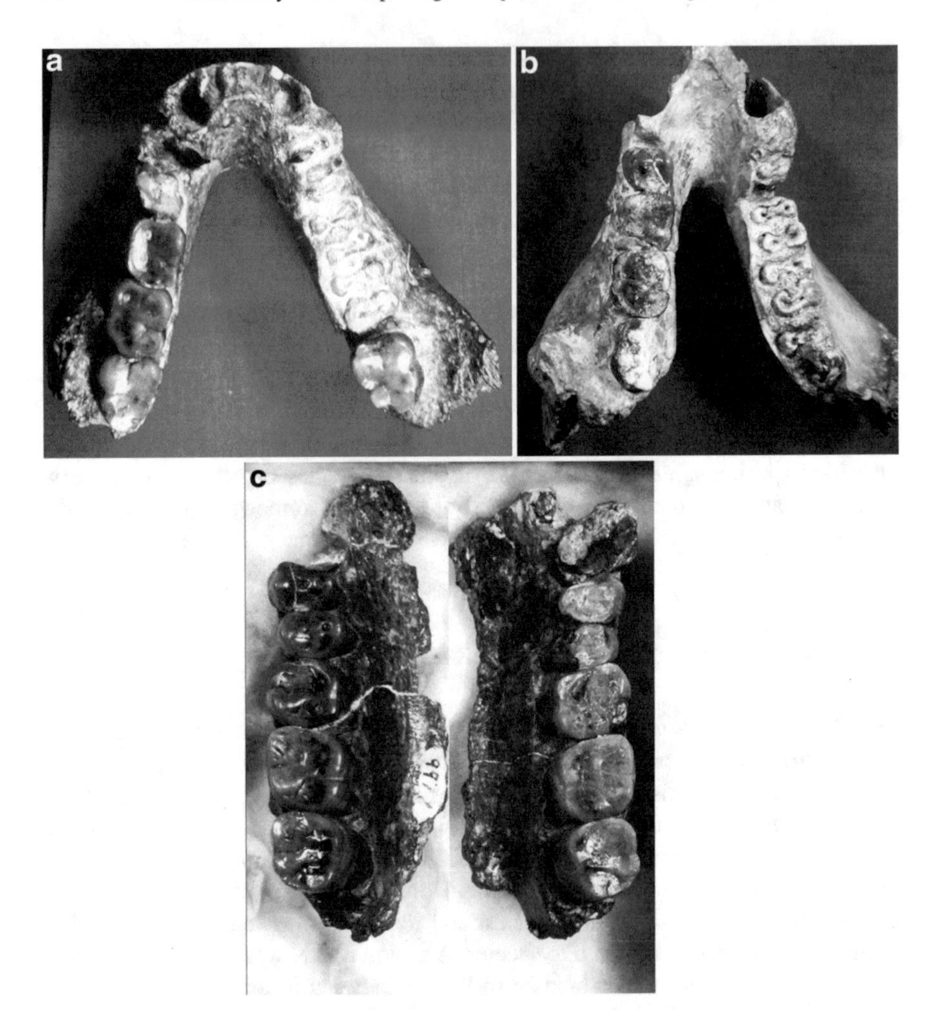

Fig. 2 Three fossils assigned to *Ramapithecus* led to a reconsideration of the genus because they do not exhibit the small canines and parabolic arcade believed to link *Ramapithecus* with humans: (**a**) GSP4857, (**b**) GSP9564, and (**c**) GSP9977

the canines, whereas the canine teeth of GSP 9977 projected conspicuously beyond the rest of the tooth row. It became difficult to separate *Ramapithecus* and *Sivapithecus* on the basis of canines. Pilbeam considered them closely related to one another even as the characters that appeared to link *Ramapithecus* with later hominins came under question as useful indicators of special relationships.

In 1977, a nearly complete face and jaws of *Sivapithecus* were discovered and received the accession number GSP 15000. Even a superficial examination of the restored face showed a remarkable similarity between *Sivapithecus* and the modern orangutan. A closer analysis confirmed this. In many small details of the face—for example, oval-shaped orbits, more vertically oriented facial bones, a general concavity of the face, and the small size of the second incisor—it is clear that the two species are close cousins. They do differ in bones of the rest of the body that were discovered later, but the relationship could not be ignored. If *Ramapithecus* closely resembled *Sivapithecus,* and *Sivapithecus* was a close relative of the orangutan, then the human link with *Ramapithecus* became untenable. Shortly thereafter, *Ramapithecus* was conceded to be female specimens of *Sivapithecus sivalensis*. Since the latter species had been named first, "Ramapithecus" ceased to be a valid name.

One issue remained to be settled. All the Asian fossils had thick enamel, like that of hominins. This included *Gigantopithecus* as well as new species discovered in China. Additional finds elsewhere shed more light on this issue. Miocene species from Hungary, Greece, and Turkey also had thick enamel. A reexamination of orangutans showed they had enamel of a medium thickness, more like modern humans and unlike the African great apes. It now appears that thick enamel was a primitive trait possessed by a common ancestor in the Middle Miocene. The gorilla and chimpanzee lineages subsequently reduced theirs. Thus enamel thickness is not an indicator of a special shared ancestry. By this time, there was no major impediment to a full acceptance of the molecular clock and the fossil tree could be redrawn.

Dissecting an Error

What went wrong? How could the fossil record have been so greatly misinterpreted? After the new discoveries were reported and models of human evolution revised, there was time for introspective soul-searching, and no one was more explicit about it than Pilbeam, who readily acknowledged and corrected his mistakes. Certainly the misunderstanding of the polarity of the enamel thickness—that is, which state was ancestral and which derived—as well as the fragmentary nature of early specimens contributed to the problem. However, these were small parts of a larger problem. Scientists operate with conceptual models of nature that guide how they interpret data, and in this case the models were wrong.

Simons and Pilbeam and others assumed that the past looked like the present. Today apes species are few, slow-breeding, and endangered. Monkey populations, on the other hand, are numerous, grow rapidly, and have spread widely across the tropics. Anthropologists were quick to assume this is the way it has always been. If there had been little diversity among fossil apes in the Miocene, it made sense that Simons had constructed his early models on a nearly one-to-one relationship between fossil and living species. Scientists now recognize that ape species multiplied quickly across Eurasia and Africa and were the dominant primates from the Early Miocene. It was the monkeys that diversified later and more slowly, emerging from Africa only in the later Miocene. Living species of apes evolved only recently and therefore cannot be linked with specific fossils from the Middle Miocene.

However, the most troubling preconception is one that has plagued philosophers and scientists through the ages and still misleads us. Humans want to think of themselves as so different from other species that they assume they have had a long separate evolutionary history. This notion has distorted anthropological thought in different ways and different times, but to Simons and Pilbeam it meant that an appearance of human ancestors as far back as 14 Ma was very reasonable. In truth, accumulating evidence from a variety of disciplines repeatedly tells us that in anatomy, genes, molecular pathways, brain structure, mentality, and behavior, the gap between humans and their nearest relative is so small that it is often hard to define.

Surprisingly, the first molecular clock study by Sarich and Wilson has not been greatly changed by the continuing stream of new and more complete genetic data. Different genes or segments of DNA have yielded slightly different dates, but there are reasons to expect this outcome. Speciation is a process that occurs over time, not an instantaneous event. Some degree of genetic variation can be expected to persist across the splitting process to the present day. Given these facts and the role of selection in speeding or slowing the rate of change, we should expect to encounter limits to the resolution of the clock. Genetic estimates for the divergence of the human lineage from that of chimpanzees now range from 4–5 Ma to 6–8 Ma. These now accord well with the fossil record, in which the earliest known putative hominins appear about 7 Ma ago.

Questions for Discussion

Q1: The molecular clock forced paleontologists to rethink and reinterpret the Miocene fossil record. Later it did the same for the evolution of modern humans. Why should we expect the molecular record to be more reliable than the fossil record? What can the fossil record tell us that the genetic studies cannot?

Q2: Sarich and Wilson calibrated their clock on the assumption that apes and Old World monkeys diverged about 30 Ma ago. More recent fossil discoveries have revised that date to the early Miocene, perhaps 25 Ma. How would that change of date affect their molecular clock?

Q3: Simons and Greenfield disagreed on whether variations among the fossils represented different species or different sexes. How should paleontologists be able to tell the difference?

Q4: Can you think of other scientists or experts in any field who publicly admitted their published interpretations had been wrong and led the way to correcting them? Why is this so uncommon?

Q5: Simons and Pilbeam assumed there were few hominoid species in the Miocene. How did that assumption mislead them?

Q6: Why is it not sufficient for scientists to correct a mistake, but also necessary to understand why it was made?

Additional Reading

Kay RF (1982) *Sivapithecus simonsi*, a new species of Miocene hominoid, with comments on the phylogenetic status of the Ramapithecinae. Int J Primatol 3(20):113–173

Pilbeam D (1980) Major trends in human evolution. In: Konigson L-K (ed) Current argument on early man. Pergamon Press, New York, pp 261–285

Sarich VM, Wilson AC (1967) Immunological time scale for hominid evolution. Science 158:1200–1203

Case Study 8. Taming the Killer Ape: The Science of Taphonomy

Abstract Hypotheses are generated within our existing understanding of the world and often incorporate societal and individual prejudices and beliefs. However, not all wrong ideas are useless: disproving hypotheses can generate new questions and hypotheses. In this example, a faulty interpretation of fossils stimulated a new field of study. From the 1930s to the 1960s, the "Killer Ape" emerged as a popular understanding of human nature as inherently violent. In this context, Raymond Dart interpreted animal bones found in caves with earliest hominins as the remains of their prey. Studies inspired by his hypotheses later proved him wrong, but challenging his ideas led to much better understanding of how fossils and assemblages are created.

Modern European prehistorians first uncovered systematic evidence of prehistoric peoples and their tools in caves in association with animal bones. They made the logical inference that these people had been hunters and were surrounded by remains of their prey. After all, hunting has a long history as a culturally important and prestigious activity and one that seems to be a link with our preagricultural past. "Man the Hunter" became an accepted part of our species identify and definition until it was challenged in the 1970s. Before then, however, Raymond Dart had given it a particularly violent interpretation. His assessment of the evidence at Taung and Makapansgat Caves proposed that our first material culture, the "Osteodontokeratic Culture" was constructed from the remains of our prey, while hunting shaped our very minds. Thus the "Killer Ape" was born.

The anatomist Raymond Dart first described and named *Australopithecus* based on the skull of a juvenile specimen from Taung Cave, South Africa. He had studied with Grafton Elliot Smith when the latter was heavily involved with reconstructing the Piltdown skull and maintained his interest in human evolution after he accepted a position at Witwatersrand University in Johannesburg. He encouraged his students to bring him any bones they came across, and thus learned of a quarrying operation at Taung that was encountering many fossils. The mine owner cooperated by sending him boxes of bones, and it was in one of these that the first skull of *Australopithecus* appeared. Dart was struck immediately by the unexpectedly large brain size and

J.H. Langdon, *The Science of Human Evolution*,
DOI 10.1007/978-3-319-41585-7_8

proclaimed it a true link between ancestral apes and humans. The initial response to his discovery from European anthropologists was skepticism—they tended to dismiss it as a fossil ape. Nonetheless, Dart persisted in his search for evidence of human origins. No more hominins have been found at Taung, but other fossils were, including a series of baboon skulls. These crania showed fractures that Dart interpreted as caused by blows of a weapon. The great majority of these were fractured on the left side, as though they had been clubbed while facing a right-handed opponent. Even a few of the australopithecine crania found elsewhere showed these injuries, hinting at murder and cannibalism.

Dart continued to collect and examine baboon remains from other South African sites, including Sterkfontein and Makapansgat Cave, to build his case that early hominins lived in the caves and were accomplished hunters. He believed the caves contained the refuse of their meals. His argument was initially based on the unlikelihood that the bones would have been accumulated by carnivores in the area, such as leopards and brown hyenas, and by his interpretation of the nature of the damage. Curiously for Dart, there were no stone tools present in these caves that might have been weapons of destruction; there were only bones of other animals.

The Osteodontokeratic Culture

Dart embarked on a detailed analysis of all 7159 fossils from Makapansgat and discovered that they were not a random accumulation of bone, but were markedly biased in favor of certain animals and body parts (Table 1). Of the vast majority of identifiable bones, 91.7 %, came from bovids (antelope) and about half of the rest were from other hoofed animals. In addition, the great majority were fragmented. Dart argued this pattern of breakage was deliberate and systematic, either through the use of the bones as tools or to shape them into more effective implements. Moreover, the edges of the fragments were smooth, as though abraded from use.

Dart concluded that the overwhelmingly most common animals, medium and small bovids, represented the preferred prey of australopithecines. Furthermore, certain body parts, when present in high frequencies, must have been valued as tools for use within the cave. When absent, they may have been removed for use outside the cave. From this, Dart proposed that the original human material culture used

Table 1 Bovid bones and bone fragments from the Makapansgat fossil deposits (from Dart 1957)

	Number	Percentage
Vertebrae and ribs	229	5.8
Upper limb	1126	28.4
Lower limb	391	9.8
Feet	864	21.8
Cranial and dental	1361	34.3
Total	3971	

tools not of stones, but of bones, teeth, and horns. He named this the Osteodontokeratic Culture. Through experiment, ethnographic example, folklore, and imagination, he pieced together their uses.

Long bones of the limbs would have made good clubs, or bludgeons, to subdue prey. When splintered, the fragments would have made blades or picks or points. Flattened pieces of bone, such as ribs, could have been used to dig or probe crevices in the rocks for food items. Jaws with the teeth still embedded would serve as serrated knives. Crania were somewhat overrepresented in the sample and may indicate headhunting for trophies, except those that were opened to access brains. The empty braincases might have been containers for fluids. Horns and bony horn cores made good pikes or picks. Even the small, roundish bones, such as astragali (from the ankle), could be projectiles. Missing body parts were either of minimal utility and never brought back to the cave (e.g., vertebrae and ribs) or may have been used in external settings. Tails, for example, might have been useful signal flags to coordinate hunters closing in on prey.

As Dart developed these ideas in scholarly publications, the depiction was well publicized in a series of books written by Robert Ardrey, beginning with the best-selling *African Genesis*. Predators, Dart believed, needed an instinct for violence. That same capacity that enabled them to kill for dinner could turn upon fellow hominins. Damage to the crania and jaws of australopithecines bore the same signs of violence as did the skull of baboons. Indeed, what difference did it make which animal was the source of their protein? Human ancestors were not only hunters, but also cannibals and possibly headhunters. They were intelligent, which made them all the more dangerous, and their inventiveness turned the bones of their prey into weapons. The blunt condyles of a humerus and sharpened slivers of a tibia were the predecessors of the mace and lance. Social predators can cooperate to make the kill, but they also can work together to defend their hunting grounds or drive neighbors away from theirs. Territoriality led to tribalism, which led to genocide. This is not a nice image in the mirror that Dart holds up, but one that was understood by a society that had experienced two world wars, the Holocaust, and the threat of nuclear obliteration. These negative perceptions of human nature took root as the Killer Ape hypothesis, which found ready acceptance in popular culture.

Dart made interesting observations and was a pioneer in his attempts to answer the questions they raised. Researching his hypothesis, he studied the behavior of hyenas and porcupines as they scavenged and gnawed bones. To understand the pattern of breakage, he considered possible butchering and exploitation strategies of the early hominins. Fortunately, others, most notably C.K. Brain, resumed where he left off and continued to improve our understanding of the story behind these bone deposits. How did the caves form? How did the bones accumulate in them? What happened to them along the way? The science that addresses these questions was given a name by Ivan Efremov in 1940: taphonomy, the laws of burial. Addressing the problems posed by Dart inspired much of the development of that field.

The Laws of Burial

Many of the famous caves with australopithecine remains lie in a small area northwest of Johannesburg now called the Cradle of Humanity. A few, including both Taung and Makapansgat, lie further away. All of these caves developed in similar ways. A cave commonly forms in a water-soluble sedimentary rock, such as limestone or (in this case) dolomite, when underground flow dissolves a cavity in the rock. As the cavity expands through further erosion, natural openings to the outside may occur to create an accessible cave, or merely a sinkhole. The subsequent history depends very much on the water levels and patterns of flow. If bones have accumulated and been buried in the caves, then they may have been part of an influx of sediment that made the cavity smaller and significantly altered its appearance.

Dart underestimated what the accumulating soil itself could do to bones. The weight of overburden can fracture and crush empty skulls while leaving those already filled with sediment undistorted. Slumping of part of the deposit can shear fossils in two. Sturdy mandibles can be squeezed together so that they break at the chin. Much of the damage to the bones that Dart interpreted as evidence of violence can be explained in this way.

Dart's observations of predators and scavengers and their handling of bones have also not held up to later research. Brain carefully documented the remains of medium-sized ungulates that have been consumed by cheetahs, leopards, and hyenas. Typically these meat-eaters focus their attentions on the trunk and proximal parts of the limbs. Cats may leave the limbs, which have little meat on them distal to the elbow and knee, relatively intact. Although the brain is highly nutritious, few animals are capable of breaking into the braincase; thus, the skull may be left alone. Hyenas are much more destructive and may consume all or most of the skeleton itself for the minerals as well as for the fatty marrow. Porcupines and smaller rodents also commonly gnaw on bones for their calcium. If the bones are not completely destroyed, the actions of predator may be recognizable by their tooth marks.

Contrary to Dart's findings, many predators do collect bones. Hyenas and porcupines have long been known to fill their dens with them, sometimes to snack upon. Leopards are known to cache their prey in trees, rock shelters, or other secluded places, out of the reach of the competition. Brain and others have found enough evidence from tooth marks on the bones at Sterkfontein, Swartkrans, and Makapansgat to implicate predators in contributing to at least part of the deposits there.

Bones that are ignored or abandoned by predators suffer further damage from weathering and trampling. Light spongy bone, including most of the bones of the trunk and spine, disintegrates more readily. Dense bones, created to withstand large forces, last longer. This would include the jaws and the weight-bearing joint areas. Most dense of all are the teeth. These properties explain the transport of the bones as well as their durability. Light bones with larger surface area are more likely to be carried downstream by a river or swept away by a flood. Dense bones or very heavy ones tend to sink and to be more quickly buried. If a skeleton is first exposed and then transported by water, one can reasonably expect bones to be sorted according to their physical properties. Any paleontologist stumbling upon them in the distant future will observe a highly skewed distribution of body parts. For these reasons, paleontologists have come to expect a preponderance of jaws and teeth. These happen to be, conveniently, highly diagnostic for identifying mammalian species.

Brain further experimented by offering a goat to a local community and then examining its bones after it was consumed. He followed this up by studying a total of 64 goat skeletons and recording the bones and fragments that resulted. His observations revealed considerable similarity to the collection from Makapansgat, suggesting that the skewed representation of bones is heavily influenced by durability of individual elements and possibly processing for food, rather than selective use of bones as tools and weapons.

Pat Shipman and Jane Phillips-Conroy performed a comparable analysis on carcasses scavenged by hyenas in Ethiopia. As at Makapansgat, the majority of the remains were of antelope. This simply represents the availability of prey to large carnivores. They found that limbs were very likely to be missing, having been carried off by carnivores or scavengers. This was more true of fore limbs since those are more easily detached. Likewise, the skulls commonly separated from the rest of the skeleton and often were transported. These missing elements are the ones most frequently found at Makapansgat. Shipman and Phillips-Conroy's conclusion is that natural actions working on the bones, including processing by predators and scavengers, can better explain the bone accumulations. Dart's evidence provides little support for hominin activity.

Fossil accumulations in other caves in South Africa are also better explained by natural processes. Today, the cavity at Sterkfontein is very much a cave in the tourist sense, when a visitor may descend fairly steeply into the depths of the rock. However, Sterkfontein is a complex fossil deposit. The richest bone-bearing part appears to have been a sinkhole opening from the top. Animals falling into this hole could be preserved relatively intact and with minimal crushing. In other parts, bones are present in the skewed proportions noted by Dart.

The cave at Swartkrans formed as a subterranean cavity that opened as the surface above it eroded. This also took the form of a sinkhole into which bones and other debris fell or were washed. It is possible that hominins once used the site as a shelter, since some tools have been found there. One dramatic discovery was an adolescent skullcap with two punctures in the parietal bones. In the same deposit was the jaw of a leopard whose canines just fit the holes (Fig. 1). Brain speculated that the australopithecine was the victim of a leopard that dragged it into a tree overhanging the cave opening. As the inedible cranium was discarded, it fell into the sinkhole to be preserved.

Taung, the original australopithecine site, contained only one hominin, a child. Although there were five small fragments representing large mammals, the great majority of bones at Taung came from much smaller ones. The culprit responsible for this accumulation was identified by characteristic scratches around the orbit of the hominin skull. This had probably been the nesting place of generations of eagles whose prey was limited by what they could carry in flight.

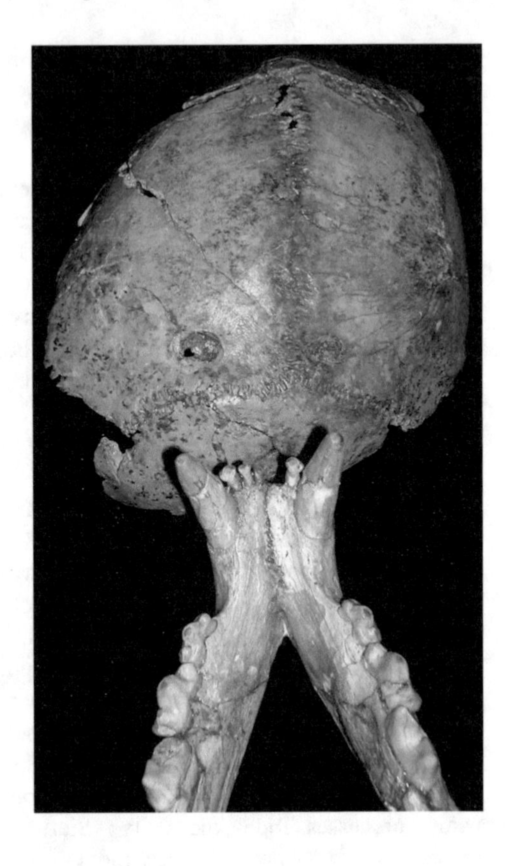

Fig. 1 This partial cranium of a young australopithecine from Swartkrans Cave bears tooth impressions that match the mandible of a leopard jaw from the same deposit. It is evidence that the bone accumulation at Swartkrans was primarily the work of carnivores, not hominins

Perspective

Did australopithecines really live in the caves? Later Neanderthals and modern peoples of Europe left evidence of tools, hearths, and artwork in caves; hence, it was natural for Dart to make this conclusion. However, not all animals whose remains are found in caves lived there. We cannot rule out the possibility that australopithecines did shelter in caves overnight or to escape the midday sun, as baboons sometimes do; yet carnivore accumulation, which is in evidence, may be sufficient to explain the bone piles. By the principle of parsimony—preference for the simplest explanation—there is no need to make such assumptions about australopithecine behavior, and such hypotheses contribute nothing to our understanding.

Victorian society was very optimistic about the potential of Western society, led by science and technology, to solve humankind's problems. The very negative view of human nature expressed by Dart and others is at least in part due to the traumas and disillusionment of the twentieth century in which technology was employed for the most destructive purposes. Fellow anthropologist Earnest Hooton echoed his perspective in 1937: "Man is a predatory mammal which has achieved dominance over all vertebrates by a ruthless use of superior intelligence." The bleak outlook on human nature reflected in the Killer Ape became a part of popular culture, inspiring authors such as William Golding, who wrote *Lord of the Flies* and *The Inheritors*.

None of the social interpretations could have revealed whether or not Dart's hypotheses were correct; those had to be tested by science. One of enduring contributions to the field is that his incorrect ideas raised interesting questions that inspired new lines of research and ultimately advanced our understanding.

Questions for Discussion

Q1: Dart published his hypothesis that australopithecines hunted baboons in 1949 based on numerous cranial remains from three caves that where both australopithecines and baboons were found. Is this a falsifiable hypothesis?

Q2: Adult male baboons are dangerous animals, more than a match for dogs, with grasping hands, intelligence, sharp teeth, and large social groups. How might an australopithecine challenge a baboon face to face or otherwise? How might australopithecines have avoided being eaten by predators themselves?

Q3: The image of humans as killer apes is bound up with many assumptions about human behavior, such as a central role for meat eating and an instinctive capacity for violence. What other assumptions or generalizations can you identify in this discussion? What cultural influences might have predisposed Dart and other anthropologists to accept these generalizations without rigorous testing?

Q4: What other kinds of evidence would more convincingly support the idea that early hominins lived in caves? That they hunted and ate other animals?

Q5: Under what circumstances can wrong ideas help rather than hinder progress in a field?

Additional Reading

Ardrey R (1961) African genesis. Dell, New York

Berger LR, Clarke RJ (1995) Eagle involvement in the accumulation of the Taung child fauna. J Hum Evol 29:275–299

Berger LR, Hilton-Barber B (2006) A guide to Sterkfontein: the cradle of humankind. New Holland, Cape Town

Brain CK (1981) The hunters or the hunted? An introduction to African cave taphonomy. University Chicago Press, Chicago

Dart RA (1957) The Osteodontokeratic culture of *Australopithecus prometheus*. Transvaal Mus Mem No 10

Dart RA (1967) Adventures with the missing link. The Institutes Press, Philadelphia

Hooton EA (1937) Apes, men and morons. GP Putnam's Sons, New York

Shipman P (1981) Life history of a fossil: An introduction to taphonomy and paleoecology. Harvard University Press, Cambridge, MA

Shipman P, Phillips-Conroy J (1977) Hominid tool-making versus carnivore scavenging. Am J Phys Anthropol 46:77–86

Tobias PV (1984) Dart, Taung and the 'missing link'. Witwatersrand University Press, Johannesburg

Case Study 9. Reading the Bones (1): Recognizing Bipedalism

Abstract How does one recognize that a newly discovered fossil is indeed a hominin? It is easy enough to list attributes that separate humans from the apes—for example, large brain, small canines, language, dexterous thumb, and paucity of body hair—but which of these appeared first and definitively in the fossil record? At the beginning of the last century, scholars like Arthur Keith mistakenly believed a large brain defined our unique lineage and tended to discount as a possible ancestor any fossil with a significantly smaller brain. Today bipedalism, habitual locomotion on two feet, is regarded as the most reliable indicator. Bipedalism is a useful trait because it represents a significant change in adaptive strategy and because there are reliable indicators in the skeleton that will be recognizable in fossils. However, as the skeleton of Lucy reveals, evidence of bipedalism does not reveal all we need to know about australopithecine locomotion. Should one focus on the differences or the similarities in anatomy? Vigorous debate in the 1980s shows that there are different ways to be bipedal.

Raymond Dart discovered *Australopithecus* in South Africa in 1924. The first specimen was a skull of a young child that possessed characteristics intermediate between those of humans and apes. Although Dart argued that its somewhat enlarged brain size and certain other traits qualified it as hominin, anthropologists in Europe hesitated to accept such a critical diagnosis for such an immature specimen. Until adult specimens became available for general examination, its status was regarded in doubt. New material was uncovered in the late 1930s and subsequently, but it was not until after the Second World War that outsiders were able to travel to South Africa to examine the fossils for themselves. Those new finds included a partial skeleton from Sterkfontein Cave, Sts 14, whose human-like pelvis convinced the profession that *Australopithecus* was indeed a hominin. Nonetheless, the incompleteness of the bones and distortion of them during fossilization made reconstruction difficult.

In this context, the discovery of Lucy in 1974 gave scientists a much more complete picture of the australopithecine body. "Lucy" was the name given to a skeleton discovered by Donald Johanson near Hadar in the Afar region of Ethiopia (Fig. 1). While it may be considered fortunate to find a partial skeleton in a South African cave where remains have lain relatively undisturbed for several million years, it is

J.H. Langdon, *The Science of Human Evolution*,
DOI 10.1007/978-3-319-41585-7_9

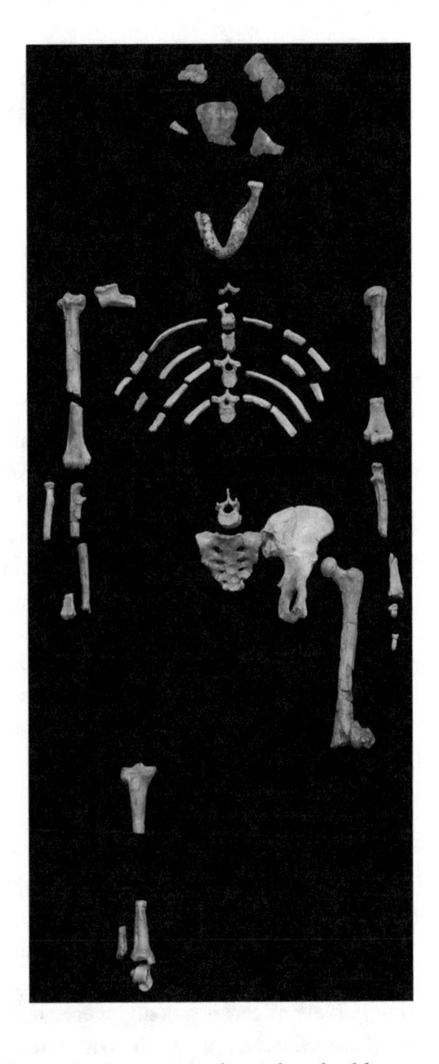

Fig. 1 AL 288-1 Australopithecus afarensis skeleton "Lucy" (cast). Source: GNU Free Documentation License, with permission

far more unusual to find more than isolated bones in the open, where they had been covered in a streambed. It was a piece of humerus that first caught Johanson's attention; then other bones began turning up. After many square meters of soil had been sifted and the bone fragments painstakingly put together, 46 % of Lucy's skeleton was represented on one side or the other. The cranium exists only as a few fragments, but we are able to study much of the rest of the body in detail. Many of the missing parts were filled in by another discovery at Hadar the following year of fragments of 13 individuals, dubbed the "First Family." They reveal a body more like that of modern humans than like apes below the waist, but with more ape-like features in the upper half. These and other fossils from Ethiopia and elsewhere in East Africa were placed in a new species, *Australopithecus afarensis*.

One of the most striking aspects of Lucy is her small stature, just over a meter (about 3.5 ft) in height. Even today she appears to be the smallest adult member of her species yet discovered. Her upper limbs suggest ape-like proportions, with long and strong limb bones, while the legs are intermediate in length. Her ribcage is also more ape-like, conical so that it is much broader at the bottom.

How Do We Recognize a Bipedal Skeleton?

The human body is reorganized from head to toe to balance and walk on two limbs. Although some of these specializations lie in soft tissues, such as the wiring of our brains and coordination of muscles, there are many indicators to be found within the skeleton (Fig. 2).

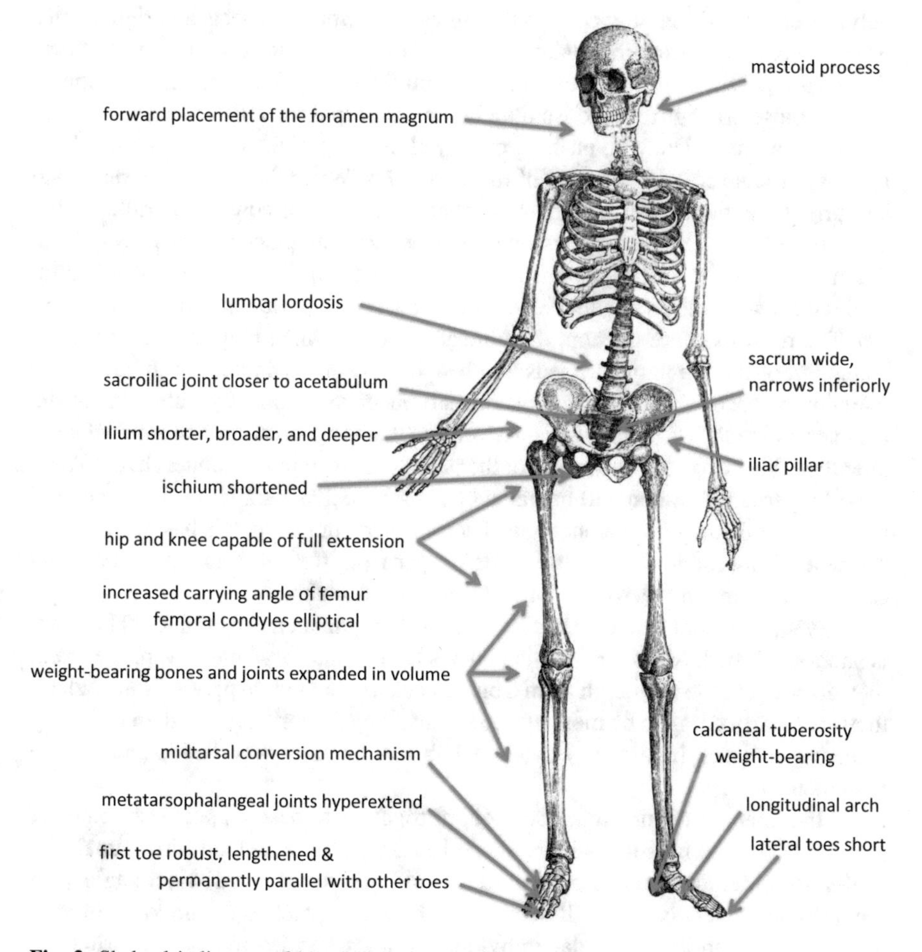

Fig. 2 Skeletal indicators of bipedalism on a human skeleton. Source: Modified from Brehms Tierleben, Small Edition 1927

The human spine is weight-bearing and there is a gradual increase in size of the vertebrae from superior to inferior. The spinal column achieved its upright position in part by bending the lower spine, creating a curvature called the lumbar lordosis. This develops as a child learns to walk and is made permanent as the lumbar vertebrae became wedge-shaped. The sacrum is wider at the top to support weight and wedges between the parts of the pelvis. Lucy's bones show this wedging. The skull needs to balance on top of the spine. The occipital condyles, where the skull contacts the first vertebra, are brought to a more central position. Muscles of the back of the neck that balance the weight of the face now attach on the underside of the skull instead of its posterior surface. These changes are seen consistently in australopithecines, although their heavy faces and jaws would have prevented the head from balancing as easily as does ours.

Perhaps the most conspicuous changes occurred in the pelvis. The mammalian pelvis is elongated and aligned with the spine. The femur intersects it at right angles when the animal is standing quadrupedally. Muscles arising from the anterior portion of the pelvis, the ilium, can pull the limb forward at the hip, in the action of flexion; those arising from the ischium behind the hip draw it back, or extend it, in the action we associate with pushing off. By elongating of the pelvis, an animal has increased the leverage and power of those muscles. When the body is reoriented to an upright position, these relationships change. The thigh now lies parallel to the axis of the spine. Hip flexors must gain power by being placed further away from that axis anteriorly, and the extensors must have an origin dorsal to it. The ventral blade of the ilium bears the attachment of iliacus, an important muscle for hip flexion. The dorsal surface anchors the gluteal muscles, which play the lesser role of abduction of the hip in quadrupeds, but become extremely important in humans to maintain balance. The iliac blade becomes broader to support a greater size of the muscles and reorients by curving anteriorly, so that the gluteal surface is now facing laterally. This brings the upper part of the pelvis into its familiar funnel shape. There is no longer any advantage to be gained by a long pelvis parallel to the spine, and there is actually a cost of balancing it. Therefore, the human pelvis has been greatly shortened from top to bottom and made deeper from front to back and, most critically, by bringing the sacroiliac joint closer to the acetabulum.

Lucy's pelvis looks much like that of a modern human at first glance. The ilium is short and broad. Although the iliac blades have some curvature toward the front, they do not have the full depth from front to back as the modern pelvis. Nonetheless, they are clearly going to be more effective balancing an upright torso than powering a quadrupedal one. In fact, they flare much farther laterally over the hip than we see in humans.

At the knee, our limbs must be brought together to best support our center of gravity. Since the hips are widely spaced to ensure an adequate birth canal, the shafts of the femurs must slant toward the midline, forming a distinct angle with the vertical. A relatively small section of the shaft attached to the knee joint is sufficient to identify a bipedal individual. Lucy's species meets that criterion. Similarly, the size of the femoral head and condyles are enlarged to distribute greater forces at the joints at both ends of the bone. Those joints are capable of full extension in us, unlike in other animals, so that we can stand without using muscular effort to support them.

The human foot has additional adaptations for walking. Because it consistently bears more weight and greater stresses, the joints are more rigidly supported and their mobility is reduced relative to what we find in a climbing ape. Instead of being thumb-like, the first toe is sturdier and longer and fixed alongside the other toes. The toes themselves are much reduced in length, having given up most of their grasping function. They are capable of hyperflexing (bending upward, as when we stand on tiptoe). The calcaneus, the heel bone, has an extra point of contact on the ground to distribute weight and improve balance. Bones of the foot in general—indeed most of the weight-bearing bones of the body—have increased in size disproportionately compared with those of other mammals to better absorb and distribute the shocks of walking and running. Lucy's foot is very incomplete, but other fossils indicate some of these adaptations were present. The first toe is larger and was held alongside the others (as also indicated by a remarkable set of 3.6 million-year-old footprints from Laetoli in Tanzania). The other toes are reduced in length, though still longer than human toes. Joints show some restrictions in range of motion and the bones are enlarged. Yet, the foot is not wholly human.

How Did Lucy walk?

How anthropologists answered this question in the past depended on whether they focused on the similarities or the differences with our own anatomy. Some of the earliest researchers still needed to argue that *Australopithecus* was our ancestor and thus emphasized indicators of human-like walking. As Johanson put it, "Here was an ape-brained little creature with a pelvis and leg bones almost identical in function with those of modern humans" (Johanson and Edey 1981). This culminated in C. Owen Lovejoy's analysis of Lucy and other remains that concluded she not only had a very human-like gait, but also differed in ways that further enhanced balance. Although skeletal differences were evident, australopithecines had clearly departed from an ape-like anatomy in a number of ways. This could not have been possible if their overall behavior and ecological niche had not also been changing. The anatomy shows the outcome of selection for bipedal standing and walking. Selection, of course, would have been more intense where the mechanical demands of efficient balance and locomotion were greater. The upper body could comfortably have lagged behind.

Critics of this approach, most notably Jack Stern and Randall Susman, asked how Lucy could have walked like modern humans if she was so different from them. The ape's upper limbs are adapted for climbing. They are long, strongly muscled and have elongated grasping hands. Stern and Susman argued that Lucy's upper body indicates climbing was still important for her. Differences from humans in the lower limb are also ape-like in direction. The bones of the toes, although intermediate in length, are curved, like those of more arboreal primates. They concluded that australopithecines still spent vital periods of time in the trees, perhaps for feeding or sleeping, and thus possessed a unique repertoire of both climbing and walking.

The two sides of this debate assumed different perspectives, from different ends of the evolutionary pathway that our ancestor was taking. It is inevitable that just as anatomy may pass through an intermediate stage not like either ancestor or descendent, so behavior and function may, too. One must also be cautious not to assume that an evolutionary path runs the straightest route.

The two interpretations made different assumptions about the chronology of evolving bipedalism. Lovejoy's model assumes that Lucy represents a snapshot of her species, a moment in time in a lineage that was constantly changing. As his colleague Bruce Latimer argued, if adaptations for climbing had been relinquished, climbing must have been less important for her than for her ancestors. And if her immediate ancestors were better climbers, her immediate descendants would surely have become better walkers. Lucy herself was not necessarily a fully adapted expert; she was merely working toward something better.

In contrast, Stern and Susman began with the assumption that any species would have to be well adapted to have survived at all. Lucy is better understood not as an undeveloped human, but as a unique and very interesting hominin to be understood on her own terms. The perspective of later discoveries supports Stern and Susman. We can now date australopithecines back to 4 Ma ago and as recently as one and a half million years ago. Although we cannot analyze their bodies through this time period in as much detail as we can Lucy's, we have to acknowledge that the australopithecine pattern of locomotion existed significantly longer than ours has been. It represents a successful strategy in its environment, not merely a way station on the road to becoming human.

Compared to living primates, Lucy is indeed unique anatomically. As anthropologists accepted this, several attempted to determine just how those differences would have affected the way she walked. Her hips were unusually wide, her toes long, and her hands swung heavily beside her thighs. It is tempting, but frustrating, to compare this to modern human gait and judge it less efficient. The difficulty lies not in our ability to discern bipedalism or describe the full range of behaviors observable in humans and apes, but in our inability to comprehend a truly different suite of movement appropriate for a different skeletal structure and ecological context.

To complicate matters, recent discoveries in East and South Africa tell us there is much more diversity of body design and locomotor patterns among the hominins than previously assumed. The *Ardipithecus ramidus* skeleton published in 2009 putatively combined terrestrial bipedalism with arboreal above-branch quadrupedalism. Skeletons of *Australopithecus sediba* (2011) and *Homo naledi* (2015) present different unique combinations of primitive and derived traits that are not simply awkward versions of modern humans. Isolated limb and foot bones from sites in East Africa and the mostly undescribed "Little Foot" skeleton from Sterkfontein are further stretching our understanding.

Questions for Discussion

Q1: Is bipedalism the most important of the characters that differentiate humans from other animals? Is "most important" the best way to select a trait to define our lineage?
Q2: Lovejoy interpreted Lucy's skeleton differently than did Stern and Susman. Could they both be correct? If not, how does one mediate such a disagreement to determine which is correct? Is either of the two models falsifiable?
Q3: Should we expect our ancestors at some point in time to show intermediate anatomy corresponding to a semi-erect posture and inefficient gait?
Q4: Is every other form of bipedalism less efficient that ours?
Q5: What might we learn from the discovery that there were many other versions of bipedal body designs?
Q6: Why has no other species, including baboons which evolved in a savanna habitat, become bipedal?

Additional Reading

Berger LR et al (2015) *Homo naledi*, a new species of the genus *Homo* from the Dinaledi Chamber, South Africa. eLife 4, e09560
Clarke RJ, Tobias PB (1995) Sterkfontein Member 2 foot bones of the oldest South African hominid. Science 269:521–524
Johanson D, Edey M (1981) Lucy: the beginnings of humankind. Simon & Schuster, New York
Langdon JH (2005) The human strategy: human anatomy in evolutionary perspective. Oxford University Press, New York
Latimer B (1991) Locomotor adaptations in *Australopithecus afarensis*: the issue of arboreality. In: Coppens Y, Senut B (eds) Origines de la Bipédie chez les Hominidés. CNRS, Paris, pp 169–176
Lovejoy CO (1988) Evolution of human walking. Sci Am 259(5):118–125
Lovejoy CO (2009) The great divide: *Ardipithecus ramidus* reveals the postcrania of our last common ancestors with African Apes. Science 326(73):100–106
Stern JT, Susman RL (1991) "Total morphological pattern" versus the "magic trait": conflicting approaches to the study of early hominid bipedalism. In: Coppens Y, Senut B (eds) Origines de la Bipédie chez les Hominidés. CNRS, Paris, pp 99–111
Susman RL et al (1984) Arboreality and bipedality in the Hadar hominids. Folia Primatol 43:113–156
Zipfel B et al (2011) The foot and ankle of *Australopithecus sediba*. Science 333:1417–1420

Case Study 10. Reading the Bones (2): Sizing Up the Ancestors

Abstract To identify a fossil mammal or to describe a new species, paleontologists like to have good specimens of the skull, especially the jaws and teeth. However, when they want to know what an animal looks like, they need to have more of the skeleton. Unfortunately, it is rare to have both skull and limb bones of the same individual, and it may be some time before scientists can reconstruct the body with some confidence. Paleontologists can apply the tools used commonly by forensic anthropologists to reconstruct stature and body proportions from individual bones to give a better picture of the size and proportions of the australopithecines and early *Homo*. The bodies of early hominins did not evolve as quickly as had been believed.

Estimating Body Size for *Australopithecus*

The estimation of stature from a skeleton or individual bones is a standard tool of forensic anthropology, where the determination of age, stature, sex, and ancestry may assist in the identification of an individual. The simplest method is to take the length of a single bone, such as the femur, and determine the correlation between that measure and body height for a population. It is not difficult to derive predictive equations for males and females. Within a measurable range of error, this method should allow us to predict stature for any member of that population. For example, Trotter and Gleser produced the following formula for calculating stature (in centimeters) in white American females from the maximum length of the femur:

$$\text{STATURE} = 2.47\,\text{FEM} + 56.60 \pm 3.72$$

This equation is entirely empirical—that is, the constants are calculated from a specific sample population and will be slightly different for any other sample. Moreover, they only apply reliably to individuals from the sampled population and in the size range of the original sample. In the above equation, the mean stature of the sample population was 160.682 cm with a standard deviation of 7.508 cm. If another

J.H. Langdon, *The Science of Human Evolution*,
DOI 10.1007/978-3-319-41585-7_10

population is substantially different in body size or composition because of age, ancestry, or nutrition, the appropriate equation will also be significantly different. Therefore all such extrapolations require caution.

The first comprehensive study of australopithecine postcranial remains was undertaken by John Robinson. He had available to him the Sts 14 skeleton and a number of isolated bones from the South African caves. The skeleton, representing *Australopithecus africanus*, included a distorted pelvis, a crushed femur, and a number of vertebrae and ribs. The pelvis and the light build of the bones in general suggested the individual was a female. Using the Trotter and Gleser equation for the length of the femur, Robinson calculated a height of 130 cm. The vertebrae were also consistent with a height between 122 and 137 cm. From the lightly built bones, he estimated a body weight of 18–27 kg.

These calculations must be put into perspective. They might be reliable if Sts 14 were a modern Euro-American female. However, *Australopithecus africanus* is clearly not a modern human, and the femur length and therefore the estimated body size were well below Trotter and Gleser's population sample. This introduces a significant, but unavoidable, degree of uncertainty into the estimate that is not encompassed by the numbers.

Robinson also had a partial humerus that seemed disproportionately long and robust for the femur. He assumed the humerus came from a male and that male australopithecines were larger than females. This is the pattern observed widely across Old World monkeys and apes and, to a lesser extent, in modern humans. Sexual dimorphism would account for only part of the discrepancy, however, and Robinson suggested the australopithecine upper limb was proportionately longer, another example of the fossil representing an intermediate form between apes and humans. Although the pelvis was unusually broad and the femur was slender, he nonetheless concluded that lower limb length was of human proportion.

The second species sampled was *Paranthropus robustus*. Parts of two femora and a distorted coxal bone were more strongly built than those of *A. africanus*. Although neither they nor other bones enabled him to calculate body height, Robinson concluded the robust species was slightly taller and substantially heavier than the gracile species. His estimate was stated as a broad range: 137–152 cm and 70–90 kg. These projections were consistent with expectations based on skulls and teeth and helped anthropologists paint a more complete picture of this phase of our ancestry while waiting for more complete material. Robinson's image of robust australopithecines was gorilla-like, and gorillas are the largest of living primates. This perception was undoubtedly influenced by the better-developed chewing apparatus, which he assumed was for a vegetarian diet, and especially by the sagittal crest atop the cranium. He therefore saw in the partial and distorted pelvis evidence of a more quadrupedal ape-like locomotion.

Additional material to improve our interpretation of these species became available over the next 20 years. By 1991, there were five partial femora of *A. africanus*. Henry McHenry applied a variety of forensic correlations to the bones and produced results similar to those of Robinson, with stature estimates of 110–142 cm. Again, one must be aware of the limitations of applying human standards to a smaller nonhuman species.

The collection of material of *P. robustus* had also grown. McHenry's observations of these bones, and also those of the robust species from East Africa, *P. boisei*, showed there to be much smaller differences in size between gracile and robust species, despite apparent large discrepancies in skull and tooth size. He estimated stature at about 132 cm for males and 110 cm for females—within the calculated range for *A. africanus*.

Calculating body mass, however, presented different issues. Because of the wide cultural range of diets and obesity among modern people, forensic anthropologists today despair of estimating body mass. Among nonhuman animals, weight for a given species varies much less. Weight should be indicated to some degree in the size of the load-bearing joints and the forces generated there. McHenry used the size of the hip joint; but apes and humans use their lower limbs—and hips in particular—in different ways. A human places full body weight on the head of the femur with each step. As is true of many human bones, the femoral head is enlarged to distribute these forces more safely. Animals that are quadrupedal distribute weight on four limbs and do not generate the peak forces that humans do as they walk and run. Thus, the equations one might use to extrapolate body mass for a human or ape are different. McHenry reported the following equations for the relationship between body weight and femoral head size (FHS):

$$\begin{aligned} &\text{for apes} && \log \text{Wt} = 2.9844 \log \text{FHS} - 2.8903 \\ &\text{for humans} && \log \text{Wt} = 1.7125 \log \text{FHS} - 1.048 \end{aligned}$$

Which equation should be applied to a fossil hominin? Choosing one over the other requires an unjustified assumption about how human-like or ape-like australopithecines might have been. McHenry conservatively chose to model the fossils twice, once assuming human mechanics and once that of apes (Table 1). His calculations, based on three specimens, ranged from 30 to 43 kg on the human scale, but 33 to 61 by ape standards. These estimates are considerably less than Robinson's estimate

Table 1 Body mass and stature of hominins scaled according to *Homo sapiens*

	Male body mass (kg)	Female body mass (kg)	Ratio M:F mass	Male stature (cm)	Female stature (cm)
A. anamensis	51	33	1.55		
A. afarensis	44.6	29.3	1.52	151	105
A. africanus	40.8	30.2	1.35	138	115
P. robustus	40.2	31.9	1.26	132	110
P. boisei	48.6	34.0	1.43	137	124
H. habilis	37	32	1.16	157	125
H. rudolfensis	60	51	1.18		
H. ergaster	66	56	1.18		
H. erectus	63	52	1.21	180	160
H. sapiens	47.9–77.8	42.4–73.2	1.06–1.24		

for *P. robustus* and much closer to those for *A. africanus*. Despite their very large teeth, the robust species was not that much larger. Robinson's implied human vs. gorilla image of contrast could be discarded.

Size Range and Sexual Dimorphism

Further insight came from a different species, *A. afarensis*, from Ethiopia. The skeleton of Lucy was described in the last chapter. Because of its relative completeness, the reconstruction of a stature of about 105 cm is more certain. While Lucy appears to be unusually small for any species of australopithecine, the "First Family" and other finds from Hadar reveal a considerable range of size. Some individuals were much larger. For example, a partial femur gave an estimate of 151 cm, half again as tall as Lucy. It is not impossible to find this range of variation among modern humans, but it would be extremely unusual to find it in a random sample of less than six individuals. Yet fossils from Hadar for many parts of the body consistently show a wide range of size. As if to confirm this, the two clearly preserved sets of footprints from Laetoli are of greatly different size.

One likely interpretation is that *Australopithecus* was a highly sexually dimorphic species. That is, males were consistently much larger than females. Modern humans and chimpanzees are mildly dimorphic, with males about 20–25 % larger in body mass than females but less than 10 % different in stature. Gorillas and orangutans are much more diverse with mature males weighing twice as much as females. Estimates for early hominins suggest a pattern of size difference closer to that of gorillas.

Uncertainties about the size of these specimens still remain. Individuals are being comparing who might have lived hundreds of thousands of years apart. In some cases, species identity is only inferred, because there is no direct association between the limb bones and cranial material that might reveal with greater certainty to which species they belong. There have been persistent suggestions that more than one species is present among the *A. afarensis* remains and also that more than one is represented among the *A. africanus* bones from Sterkfontein Cave in South Africa.

Gender must also be inferred. If it is assumed that all large individuals are male and all small ones are female, this will improperly confirm the hypothesis with circular reasoning and certainly exaggerate the actual dimorphism. It is usually difficult to identify the sex of most modern human skeletons from the bones alone. The pelvis is the most reliable indicator—but not a perfect one—because of the relationship between the shape of the pelvis and the birth canal for a large-brained infant. Ape infants have small brains and birth is less difficult. Aside from gross size, there are few differences in the pelves of male and female apes. Instead, the development of the attachment areas on the skull for chewing muscles and the length of the canine tooth are much more reliable indicators of sex in monkey and ape skeletons. To identify the sex of australopithecines, should one examine the pelvis or the skull? The truth is, only two examples of pelves are known that are reasonably complete and have been well described. Both are believed by most researchers to be female, but it is unlikely that the ape-size brain of australopithecines would have required much adaptation in their mother's pelves. Thus males and female bones may not differ very much. On the other hand, differences in the robusticity of the skulls, including conspicuous crests for muscle attachment in some specimens, suggest an ape-like pattern of dimorphism.

The implications of high levels of sexual dimorphism are interesting, but highly speculative. The most common explanation of dimorphism in primates is sexual competition among males. Larger and stronger males are more likely to reproduce, either because females select them or because they defeat or intimidate their rivals. When males are large, we usually assume that a few successful males can monopolize a much larger number of females and that many males are shut out of mating opportunities. On this basis, it has been argued that australopithecines may have had a mating system like that of gorillas, in which one male dominates a "harem" of females until he is overthrown, or like that of baboons, in which a core of mature males hold power and the attention of most of the females in a much larger social group. Such a social structure has been proposed by Charles Lockwood for *P. robustus* on the basis of cranial dimorphism.

Later hominins—members of *Homo*—seem to show a reduced level of dimorphism. This might indicate a different social structure. In modern human societies, men and women commonly form pair bonds to maintain a household and raise children together, thus making opportunities are more evenly shared among males. When did our modern mating structure arise? The skeletons provide the only basis for such speculation.

Primitive Body Proportions

The cave at Sterkfontein, where most of the *A. africanus* material was found, has continued to produce more fossils. McHenry and Berger attempted to sort these into large, medium, and small body sizes and found a surprising result. Nearly all of the upper limb material (22 of 23 specimens) appears to come from medium or large individuals. In contrast, 25 lower limb specimens were classified as small and only three were put in the medium size category. One partial skeleton (Sts 431) was considered large-bodied for the upper limb and medium for the lower limb. The only reasonable explanation is that the upper limb bones of *A. africanus* are proportionately longer and more robust than those of the lower limb—more so than for modern humans and more so than for *A. afarensis*.

Humans have greatly elongated lower limbs relative to other primates. The australopithecine pattern is interpreted as primitive. Long and strong arms are important in climbing for modern great apes and, presumably, for *A. africanus*. Scientists are faced with a discordant image of an animal with an apelike upper body that stands upright on relatively short legs. Contrary to previous assumptions, it also appears that the two best-known australopithecine species, *A. afarensis* and *A. africanus*, differed in their body form and probably in the way they used the environment.

Early *Homo*

Another surprise came from the earliest *Homo*. In 1986, Donald Johanson and Tim White announced the discovery of a highly fragmentary skeleton from Bed I of Olduvai. Small pieces of bone and tooth were scattered across many meters on the floor of the gorge. The field crew had to sift immense amount of dirt and painstakingly piece the bones together. The new specimen, designated OH 62, includes parts of the limb bones and enough of the teeth to allow identification as *Homo habilis*. The shafts of the long bones, missing the joint surfaces, have been the subjects of many attempts to reconstruct body size and limb proportions. The right upper limb bones—humerus, radius, and ulna—are mostly present, but the lower limb is represented only by the proximal part of the left femur, missing the head, and a small piece of proximal tibia. Uncertainties of reconstruction increase the controversy over their interpretation.

The OH 62 femur is more slender and lightly built than that of Lucy. Some researchers therefore reconstructed it with a shorter estimated length. This produces the surprising result that like *A. africanus*, *H. habilis* had a lower limb that was proportionately shorter compared to the upper limb than did *A. afarensis*. However, others have suggested another reconstruction. They compared the OH 62 femur to that of another bone found at Olduvai, OH 34, which had not been assigned to a species. OH 34 has a similar slender build, but is much longer than an australopithecine femur. If OH 62 is regarded as similar to OH 34, the lower limb size falls more comfortably into the range of modern humans, though the upper limbs are still long. *H. habilis* would then appear significantly more derived compared to Lucy. However, the species identification of OH 34 remains uncertain. This femur comes from Bed III, dated to 1.15–0.8 Mya, much later than any other known *H. habilis* fossils. The other hominins known from that period at Olduvai are *H. erectus* and *P. boisei*. Both species have more robust bones and specific features that make OH 34 a poor match. If OH 34 is not a very late surviving *H. habilis*, perhaps it is an extreme variation of *H. erectus*.

Does it matter? What are the implications of these different reconstructions? Most paleoanthropologists consider *A. afarensis* to be the best candidate for a direct human ancestor about 3.5–3.0 Mya and *A. africanus* to be a contemporary or slightly later geographical variant from the south. *H. habilis*, as the most primitive member of *Homo*, is often presented as another possible direct ancestor. This comfortable picture is disturbed if one view the limb proportions of *A. afarensis* as more humanlike than those of either of the other species. One must either argue for an evolutionary reversal or dislodge *H. habilis* from our lineage. Indeed, largely on the basis of more primitive limbs Bernard Wood and Mark Collard argued to remove this species from *Homo* and put it into *Australopithecus*. If, on the other hand, *H. habilis* had limb proportions consistent with other members of *Homo*, the species would rest more comfortably where it is.

This century has witnessed a remarkable series of discoveries of partial skeletons of additional species of both *Australopithecus* and *Homo*. Body mass estimates have been summarized on Table 2. Unfortunately, the material does not permit a valid estimation of the level of sexual dimorphism, but these new specimens increase the range of known morphology and expected body size.

Table 2 Body mass estimates for recent hominin discoveries

	Sample	Mass (kg)	Stature (cm)
Australopithecus sediba	$n=1$	32–36	
Homo cf. erectus (Dmanisi)	$n=2$	40–50	145–166
Homo floresiensis	$n=1$	16–29	106
Homo naledi	$n=4$	40–56	145–149

Questions for Discussion

Q1: In order to estimate stature from femoral length, what assumptions are being made about the australopithecines?
Q2: Why would extrapolating stature estimations to individuals outside the range of the original sample populations be unreliable?
Q3: The standards for estimating body mass derived from living apes may or may not be more appropriate for australopithecines than human standards. Which do you think might be more appropriate, and why?
Q4: What other adaptive advantages, besides male competition, might there be for different body sizes in males and females?
Q5: If *H. habilis* is descended from *Australopithecus* and is ancestral to later species of *Homo*, how do we decide to which genus it belongs?

Additional Reading

Anton SC (2003) The natural history of *Homo erectus*. Yrbk Phys Anthropol 46:126–170
Anton S et al (2014) Evolution of early *Homo*: an integrated biological perspective. Science 345:45
Berger LR et al (2015) *Homo naledi*, a new species of the genus *Homo* from the Dinaledi Chamber, South Africa. eLife 4, e09560
Brown P et al (2004) A new small-bodied hominin from the Late Pleistocene of Flores, Indonesia. Nature 431:1055–1061
Haeusler M, McHenry HM (2002) Body proportions of *Homo habilis* reviewed. J Hum Evol 46:433–465
Hartwig-Scherer S, Martin RD (1991) Was “Lucy” more human than her “child”? Observations on early homind postcranial skeletons. J Hum Evol 21:439–449
Lockwood CA et al (2007) Extended male growth in a fossil hominin species. Science 318:1443–1446
McHenry HM (1986) Size variation in the postcranium of *Australopithecus afarensis* and extant species of hominoidea. In: Pickford M, Chiarelli B (eds) Sexual dimorphism in living and fossil primates. Il Sedicesimo, Firenze, pp 183–189
McHenry HM (1991a) Femoral lengths and stature in Plio-Pleistocene hominids. Am J Phys Anthropol 85:149–158
McHenry HM (1991b) Petite bodies of the “robust” australopithecines. Am J Phys Anthropol 86:445–454
McHenry HM, Berger LR (1998) Body proportions in *Australopithecus afarensis* and *A. africanus* and the origin of the genus *Homo*. J Hum Evol 35:1–22
Richmond BG et al (2002) Early hominin limb proportions. J Hum Evol 43:529–548
Robinson JT (1972) Early hominid posture and locomotion. University Chicago Press, Chicago
Trotter M, Gleser GC (1952) Estimation of stature from long bones of American Whites and Negroes. Am J Phys Anthropol 10:463–514

Case Study 11. The *Habilis* Workbench: Experimental Archaeology

Abstract The earliest stone tools were crude: a hominin picked up a cobble and bashed something with it. If the cobble broke, it produced a sharp edge, opening up further possibilities. Over the course of a million years or so, hominins became increasingly skilled in shaping the tools with relatively few flakes so they would be appropriate for the task at hand. Such tools must lie by the millions across the African landscape, or still buried, waiting to be found. Mary Leakey found them in great quantity at Olduvai Gorge and in the tradition of archeologists, she catalogued and described them according to shape. Functions of the tools and the actions of the tool-makers were left to the imagination until a new generation of researchers brought experimental archaeology to the field and shed new light on life in the very earliest Paleolithic era.

The Oldowan Tools

Beds I and II at Olduvai reach from nearly 1.8 to 1.2 Ma ago. For some 600,000 years, hominins made and abandoned their artifacts with only the most gradual improvements in sophistication. Elsewhere in East Africa, especially in Ethiopia, similar stone tools go back even further, to nearly 2.6 Ma, and they are no different. One should not assume that this represents the start of tool use by hominins. Even more crude tools dating to 3.3 Ma were announced in 2015. Chimpanzees and other primates make and use a variety of tools and one may assume that the common ancestor had similar technologies. Of these, however, the only durable tools are hammer stones and anvils used by chimps and capuchin monkeys to crack nuts. Nonetheless, something significant changed by 2.5 Ma, because stone tools become increasingly frequent and widespread.

Mary Leakey was responsible for the systematic description and analysis of the tools from Olduvai. In the traditional approach of archaeologists at the time, she painstakingly sorted and named them by shape (Table 1). "Heavy-duty" tools were

J.H. Langdon, *The Science of Human Evolution*,
DOI 10.1007/978-3-319-41585-7_11

Table 1 Mary Leakey's classification of Oldowan artifacts, based on shape with inferences about possible use (from Leakey 1971, 1974)

Tools	Choppers	Side choppers
		End choppers
		Two edged
		Pointed choppers
		Chisel-edged
	Proto-bifaces (rare)	
	Bifaces	Irregular ovates
		Trihedral
		Double-pointed
		Flat
		Cleavers
		Oblong picks
		Heavy-duty picks
	Polyhedrons	
	Discoids	
	Spheroids	
	Subspheroids	
	Modified battered nodules and blocks (last 3 categories blend into one another)	
	Scrapers	End
		Side
		Discoidal
		Perimetal
		Nosed
		Hollow
	Burins	
	Awls (developed Oldowan only)	
Utilized material	Anvils	
	Hammerstones	
	Cobblestones and nodules	
	Utilized flakes	Straight
		Convex edge
		Concave edge
Debitage	Flakes	Divergent
		Convergent
		Parallel-sided
	Resharpening flakes (from resharpening choppers)	
	Broken flakes, impossible to classify	
	Core fragments	
Manuports		

Fig. 1 Oldowan chopper from Olduvai Gorge, 1.8–2.0 Ma. Source: Creative Commons, with permission

the cores of cobbles from which flakes have been removed. The most common of these were choppers, which have at least one sharp edge (Fig. 1). Other terminology refers more explicitly to the shapes of the tools—polyhedron, discoid, and spheroid—or to the position of the working edge—side chopper, pointed chopper, and two-edged chopper. A second category of tool is made by modifying, or retouching, the flakes that came off a core. These light-duty tools may be described by names suggesting possible functions, such as scrapers and burins. A third category of "utilized" stones refers to material that was unmodified except for damage resulting from use. These include stones used as hammers or anvils. In addition, Leakey noted debitage, flakes produced as by-products during tool manufacture, and manuports, stones that apparently were carried into the site from another location. Manuports might have been raw materials that were never utilized.

She further recognized some changes in the tools over time. In Bed I, the lowest in the sequence, there was a predominance of heavy-duty core tools. This was the classic Oldowan Culture. In Bed II, there was a transition to a greater frequency of small flake tools and spheroids. She called this the Developed Oldowan. (Also in Bed II another culture appeared, the Acheulean tradition, which is recognized as still a later and more sophisticated technology.)

Louis and Mary Leakey developed a systematic excavation approach to the Oldowan sites at Olduvai. Instead of merely collecting tools, they exposed large horizontal surfaces and recorded the relationships of tools and bones. Careful mapping of these past land surfaces showed the distribution was not random. In several instances, fossilized animal bones and artifacts were concentrated in rough circular patterns, indicating activity areas. At one site in particular, designated DK, there was a circle of unmodified stones that Leakey interpreted as the base of a shelter.

The stones might have anchored branches or some perishable material forming the type of hut used by some hunter-gatherers in Africa in modern times. Other, natural interpretations have been offered, and the proper explanation may never be known.

Despite Leakey's careful work, there was only speculation concerning the uses to which the tools were put. Terms such as "scrapers" and "choppers" are defined by shape alone. The names suggest uses for which the tools could have been employed, but there was no direct evidence behind them. There was a clear inference, since both tools and animal bones occupied the same levels and often the same spaces, that hunting and butchering were important activities. Heavy choppers would make effective weapons at close range. The round spheroids were suggested by Louis Leakey to be bolas weights. Some of the animal bones were broken in ways she interpreted as deliberate shaping or wear resulting from their use as tools.

Who made the tools? In 1959, when the only hominin known from Olduvai was *Zinjanthropus* (now *Paranthropus boisei*). Louis Leakey credited this robust australopithecine with them. Two years later, when he had named *Homo habilis,* also from the Gorge, Leakey ascribed the tools to *H. habilis* and the remains of *Zinjanthropus* to the habiline dinner table. He reasoned that the species most closely ancestral to us was most likely the toolmaker. Of course we cannot know who made the tools. Perhaps both did, or neither, since we now know there were other hominin species present in East Africa at this time.

Experimentation

Archaeologists of a new generation, such as Kathy Schick and Nick Toth, endeavored to get a better sense of the minds that made the Oldowan tools. Experimental archaeology reproduces the behavioral processes by which a site or artifact was created. It is more interested in how a tool was created and used than in what it looked like. The art of making stone tools has been rediscovered many times by archaeologists and occasional forgers. Trial and error informs researchers of the preferred raw materials—fine-grained, hard stone that fractures in any plane. Not surprisingly, modern-day knappers and early hominins agree on those preferences.

A replicator of Oldowan tools begins with a mental image of the tool he is trying to copy, then strikes flakes to shape the core accordingly. Did Oldowan artisans also have a mental image to follow? This question probes at a possible watershed between ape and human minds. Some scholars, such as Thomas Wynn and William McGrew, argue that the mental steps necessary to make Oldowan tools are within the capability of chimpanzees; others, such as Anne Delagnes and Helene Roche, suggest that considerable planning and intentionality of form were present. Unfortunately, attempts to test this directly by teaching chimpanzees to make stone tools are frustrated by physical limitations of their handgrip. Their thumbs are simply too short and lack the same degree of motor control that humans have. One particularly intelligent and cooperative subject, a bonobo named Kanzi, initially found it as effective to throw the rock against the cement floor as to strike it with a hammerstone.

Modern experimentation showed that some foresight was needed, if only to have the proper raw material on hand. The act of flaking a tool could be as simple as placing it on an anvil stone on the ground and striking it with a hammer stone. This is the same technique used by chimpanzees to crack nuts and would account for the many "utilized pieces" in Leakey's collection. Edges of flakes or cores could be used to cut and carve wood. This enabled Schick and Toth to make a crucial tool used around the world, a digging stick. Especially in drier country, such as the African savanna, many plants put their most desirable resources underground in roots and tubers where few animals have access to them. A digging stick, therefore, makes a wider range of food resources available. The stone tools were also appropriate for scraping a hide, although other steps, such as tanning, would have been necessary before the leather could be very useful. Another task investigated was butchering prey animals. Schick and Toth successfully dismembered and harvested meat from a number of domestic and wild animals, including both a wildebeest and an elephant.

Putting the tools to use caused them to rethink the meaning of Oldowan tools. Although Mary Leakey cataloged retouched flakes and recognized the potential use of unmodified flakes, she concentrated on the cores. In practice, the heavy core tools were adequate for working wood, butchering animals, or smashing bones. However, the flakes provided good sharp edges, as well. Tools would frequently become dull, break, or be lost in the carcass. It was far easier to strike off a new flake than to resharpen the old one or to make a new core tool. Many of the cores, therefore, may be the by-products of making flakes.

Clearly, core tools were also intentionally created. Some tasks require a stone with a larger handgrip or more weight to it. In these cases, foresight and a mental image are needed, and the flakes are removed in a predictable sequence. By piecing the core and flakes back together into the original cobble, Toth was able to reconstruct the steps taken toward the final result. This was actually possible with some Oldowan tools where the flakes were recovered where they had been left 2 Ma before. In some cases, the debitage flakes were left on the ground, but the core had been carried off. It was possible for Toth to discover the shape of the missing core by putting the flakes together into a negative mold. Thus, the experiments allowed him to distinguish between tool and by-product and to reconstruct the intent that determined a methodical sequence of flake removal. He was even able to argue that Oldowan knappers were predominantly right-handed.

Spheroids had also been misunderstood. Mary Leakey recognized a continuity of form among polyhedrons, subspheroids, and spheroids, suggesting that polyhedrons and subspheroids were early stages in the production of a spheroid. Toth found that a hammer stone used repeatedly becomes increasingly round. The tool-user will direct the most projecting part of the stone against the anvil to deliver the greatest pressure. The result is that the high points are chipped and worn down until the hammer resembles a sphere, a by-product of the activity.

Manuports

An experienced tool knapper would select his material carefully. Many types of stones are too soft or flake poorly. Learning what materials Oldowan toolmakers used and where they obtained them may give some insight into the degree of their insight and foresight. Fortunately, the harder rocks suitable for tools also have some likelihood of surviving erosion and persist today. If those sources of raw materials remain exposed, the opportunities available to early hominins can be appraised. In a few cases, it is possible to trace a tool to its exact source because of its precise chemical composition. At Olduvai, the source of lava used as a preferred material was 2–3 km from the place the tools were discovered. At other sites, evidence is less clear, but suggests similar distances.

Hominins who carry stones or tools 2 or 3 km are showing both appreciation of stone quality and foresight in anticipating their needs. It would be inefficient to walk kilometers to fetch a single tool when needed, and equally so to carry heavy raw materials with them "just in case." However, if a hominin placed a supply of the stones at a location he/she could remember, they would be more readily available. Such caches have been discovered far from their sources. They might have been deliberately placed or casually abandoned after use. Even a single core that could be used to produce flakes would be a resource to which a hominin might bring small prey or part of a carcass for processing.

Home Bases

Mary Leakey's unique discovery of a ring of stones is suggestive, but not at all conclusive, evidence of a shelter. She was drawing on ethnographic analogy—the resemblance of the circle to the arrangement of stones created by some modern peoples for shelters. However, making such an analogy requires the assumption that early hominins at that time were behaving like modern hunter-gatherers in building shelters at all. Such an assumption was made more explicit in the critiques of a model put forth by archaeologist Glynn Isaac.

Isaac distinguished three types of sites he designated Types A, B, and C. Type A sites consisted of an assemblage of artifacts: tools, debitage, and manuports. Type B sites had the bones of a single animal associated with artifacts. Type C sites contained bones of a number of different animals plus artifacts. These he related to activities in a seemingly straightforward manner. Type A sites were simply places where hominins paused to make tools. Type B sites represent the remains of a single kill, where hominins had gathered to butcher an animal before moving on. Type C sites were home bases, camps where hominins slept, where they tended children, from which they set out in the morning to hunt and gather, and to which they returned with food to share. Isaac was explicitly comparing the supposed behavior of early hominins to the observed patterns of living of modern hunter-gatherers, such as the well-studied !Kung Bushmen. Home bases and food sharing are universal behaviors of modern humans. Was he justified in those assumptions? Were the early hominins human in this sense?

Although a few primates cache their young, leaving them in a secure place while the adults forage, most primates and all great apes are more mobile. Each night they are likely to sleep in a different place than the night before. They may return periodically to a preferred grove of trees, but only as stops on larger circuits through their territories. Chimps and gorillas build nests, but only for use for a single night. It might be asked when and why humans began to occupy home bases. "Why" likely relates to increasing dependency of infants and children, which in turn relates to economic sharing of food. Dependency is probably a correlate of increasing brain size, and thus is more likely to have developed at a later date. There is little reason to model Oldowan hominins after humans rather than apes.

Other explanations could account for Isaac's three types of sites. Type A sites might have resulted from working with perishable materials such as wood, or with bones that were later removed by scavengers. Type B sites are likely to be evidence of butchering activities, but even hunting is an inference that is challenged in the next case study. Type C sites involve an accumulation of material over time. Both carnivores and hominins are capable of accumulating bones by moving them to a central location, but water can do this also. There are circumstances, such as drought or a natural trap that may cause several animals to die at one place. If one does assume that butchering and or consumption was occurring at that site, how can one differentiate between a favored picnic spot and a home base? Perhaps the spot was preferred because of other nearby resources, such as shade, water, raw materials for tools, or a good view of approaching predators. Many other assumptions need to be reconsidered. Were the hominins sleeping there? Did child care occur more often there than elsewhere? Was occupation continuous or intermittent? Were individuals integrated into mutually dependent economic groups? Was there a division of labor? Was food being brought to others? Was it shared? Until those questions can be addressed, it is presumptive and dangerous to leap to conclusions about a home base.

One predicted difference between living sites and repeated use locations is the temporal pattern of use. Modern hunter-gatherers tend to stay in one location until the local resources are tapped out, then the band will relocate. Refuse thus accumulates for weeks or months only. In contrast, at a site visited only occasionally, bones might accumulate over a longer period of time, perhaps years. Observations of the fossils sometimes show bones in many different stages of weathering from exposure to sun and rain. Although some have suggested this is evidence for intermittent visitations, rather than a single episode, the same observations may also be explained if some of the bones became buried sooner than others.

Overall, there is evidence for a wide range of behavioral patterns. Raw materials were carried significant distances. Cores and/or flakes were frequently carried, as they are not often found at the same sites. Finally, there is evidence that sites were used repeatedly, but very possibly discontinuously. This tells us that not all tool-making and tool-using stages were carried out in the same place.

The Oldowan tradition is the oldest material culture for which there is extensive direct evidence. Whether or not other hominins were making these tools, at least some of these sites can be associated with early *Homo*. Experimental studies of Oldowan tools give a partial glimpse into the capabilities of the human mind at this early period and provide evidence of planning and intention, but very little of innovation.

Questions for Discussion

Q1: Archaeologists in this study deliberately made tools to look like Oldowan tools. Did Oldowan toolmakers make tools according to the way they looked? Does it matter for our understanding of the tools?

Q2: A screwdriver has a specific function, of turning screws, but how many other uses do we have for it? Why is it even useful for us to ask the function of an Oldowan tool? Can we ever know?

Q3: A modern toolkit in our society may or may not look like a modern toolkit used in China or Africa, because they may design tools differently or they may have different uses for tools. Should we expect similar diversity of toolkits in different prehistoric populations? Would we be able to recognize such differences if we found them?

Q4: Could people survive in a natural habitat without tools like most animals do?

Q5: Why do human societies have home bases while other primate species do not need them?

Q6: Why should we assign the Oldowan tools at Olduvai to *Homo habilis* or another species of early *Homo*? What would be the implications if a different species made them?

Additional Reading

Delagnes C, Roche H (2005) Late Pliocene hominid knapping skills: the case of Lokalalei 2C, West Turkana, Kenya. J Hum Evol 48:435–472

Harmand S et al (2015) 3.3 Million-year-old stone tools from Lomekwi 3, West Turkana, Kenya. Nature 521:310–315

Isaac G (1978) The food-sharing behavior of proto-hominids. Sci Am 238(4):90–108

Leakey MD (1971) Olduvai Gorge, vol 3, Excavations in Beds I and II, 1960–1963. Cambridge University Press, Cambridge

Leakey MD (1974) Preliminary survey of the cultural material from Beds I and II, Olduvai Gorge, Tanzania. In: Clark JD, Bishop WW (eds) Background to evolution in Africa. University Chicago Press, Chicago, pp 417–442

McPherron SP et al (2010) Evidence for stone-tool-assisted consumption of animal tissues before 3.39 million years ago at Dikika, Ethiopia. Nature 466:857–860

Plummer T (2004) Flaked stones and old bones: biological and cultural evolution at the dawn on technology. Yrbk Phys Anthropol 47:118–164

Schick KD, Toth N (1993) Making silent stones speak. Simon & Schuster, New York

Toth N, Schick K (eds) (2006) The Oldowan: case studies into the earliest Stone Age. Stone Age Institute Press, Gosport

Wynn T, McGrew WC (1990) An ape's view of the Oldowan. Man 24:383–398

Case Study 12. Hunting for Predators: The Scavenging Hypothesis

Abstract Difficult questions often spur the imagination to find new lines of inquiry. Even if it cannot answer basic questions, such research may produce surprising ideas and new perspectives. Dart's Osteodontokeratic culture was put to rest by the science of taphonomy, but questions of prehistoric carnivory remained. Dart and Ardrey had depicted human ancestors as innate predators and meat-eaters. Even discounting the lurid images conjured by Killer Apes, "Man the Hunter" was the foundation of models of human evolution in the mid-century. This was subsequently challenged for political reasons, and tested by scientific methods. Over the succeeding decades, competing models of "Woman the Gatherer" and "Man the Scavenger" demanded consideration. A more sophisticated—and balanced—understanding emerged with the introduction of new tools, including applications of the electron microscope. One site where this issue has been debated is FLK Zinj in Bed I of Olduvai Gorge, where the original cranium of *Zinjanthropus* was found. The identification of tool cut marks on bones found there confirmed that hominins were processing carcasses, but so were carnivores. Anthropologists were forced to take a closer look at our ancestors' position in the ecosystem, and consider a broader interpretation.

The diet of a species, living or fossil, relates to many dimensions of its niche, including habitat, land use, locomotion, anatomy, social behavior, and life history strategy. The ancestral human diet has been the focal point of many scenarios about human evolution. Different researchers have hypothesized meat, bones, shellfish, tubers, seeds, and fruit as the staples of that diet, which determined the evolution of our most distinguishing characteristics. The ancestral, or "Paleo" diet has attracted interest in medicine and marketing today.

Diet is also an abstract concept. It is one thing to list the food items a person ate yesterday. It is another to generalize that to a lifetime, much less to a species. It is difficult to know which food items would relate to natural selection, particularly in an omnivore. Moreover, it is desirable to sort species into categories so that we can compare them and identify meaningful correlates of anatomy, behavior, and ecology. Generalized terms such as frugivore (fruit-eating), folivore (leaf-eating), insectivore

J.H. Langdon, *The Science of Human Evolution*,
DOI 10.1007/978-3-319-41585-7_12

(insect-eating), or carnivore (flesh-eating) do not describe whole diets. Humans are considered omnivorous because they regularly consume foods from all these categories and more. It is clear that in a strict sense all monkeys and apes, as well as many other mammals, are omnivorous; but what is the value of such a label?

What questions are really being asked when by an attempt to reconstruct early hominin diets? Usually one is looking for specializations that may help explain adaptive behaviors and anatomical features. According to the data described above, later australopithecines may have evolved such specializations not by changing broad dietary categories, but by shifting to the tougher foods of a drier habitat. To the extent they may have incorporated additional amounts of meat in the diet, more recent hominins would have become more general, rather than more specialized. As omnivores, hominins may define dietary generalization as a reduction of specialization. Generalists are, by definition, less likely to display anatomical structures and behavior patterns associated with any particular dietary category, and it is unlikely that "human nature" will be explained by a specific diet. Nonetheless, omnivory does not mean a lack of discrimination. The human diet is opportunistic and highly selective to assure an efficient return for foraging effort.

The Diet of our Ancestors

As noted in previous chapters, anthropologists began with a prejudice that focused on hunting as a defining aspect of our behavior. For example, John Robinson proposed the Dietary Model to explain the very small differences between the dentition of *Australopithecus africanus* and *Paranthropus robustus*, the first a tool-using hunter and the latter a gorilla-like vegetarian. However, upon a more objective look at the fossils, analogies with living apes or other mammals cannot support such a distinction.

In 1970, Clifford Jolly proposed the Seed-eating Hypothesis. Building on an analogy with gelada baboons, he argued that eating grass seeds could explain the evolution of hypertrophy of australopithecine molars, canine reduction, finger dexterity, and bipedalism. This was a departure from previous ways of thinking about human ancestors and stimulated the questioning of previous assumptions. His hypothesis was critically tested by examining the enamel surfaces of hominin teeth under an electron microscope. At that magnification, wear patterns more closely matched a fruit-eating diet than either seed- or meat-eating.

The scanning electron microscope (SEM) was again applied to ask the question whether early hominins ate meat at all. While Louis Leakey and others assumed that the numerous animal bones found at the same sites as australopithecines and early *Homo* were the victims of hunting, there was little direct evidence for this. Two studies, by Henry Bunn and by Richard Potts and Pat Shipman attempted to identify cut marks made by stone tools on these bones. They were published as adjacent articles in *Nature* in 1981.

Bunn examined bone fragments from FLK *Zinj* at Olduvai Gorge in Tanzania and two collections at Koobi Fora in neighboring Kenya. He distinguished damage cause by tools, including both cutting and hammering activities; teeth of carnivores and rodents; and weathering and other wear to which the bones would have been subject over the past 1.5–2 Ma. Bunn found butchery marks present on about 300 bones from Olduvai and about 20 specimens from Koobi Fora. These bones represent a wide range of horse, pig, and antelope species. Others bore percussion marks from where they had been smashed open to access marrow. Bunn's findings provided strong evidence that hominins were consuming meat.

Potts and Shipman focused their study on bones from multiple sites at Olduvai, including FLK *Zinj*. They also observed both tooth and tool marks, with more of the former. Interestingly, several specimens showed both types of damage, suggesting hominins and carnivores were in competition for the carcasses. Moreover, there was some differentiation of the body parts affected. Three quarters of the surfaces with tooth marks were on bones from meat-bearing parts of the animal, whereas only half of the tool marks were. The rest occurred on bones from the ends of the limbs, where there were tendons, but no flesh.

These studies provided gratifying confirmation of the assumption that hominins were eating meat, but they posed additional questions and possible interpretations. If both hominins and carnivores processed the same bones, who had them first? Who was the predator and who was the scavenger? Shipman expanded her study to try to answer this question. Sorting through over 2500 antelope bones, she found 13 examples where tooth marks intersected tool marks. In eight of those, the carnivore was there first. Compared with patterns at a more recent Neolithic site, the Pleistocene bones bore a smaller percentage of cut marks from disarticulating the joints. Disarticulation would be important if the hominins were cutting up the carcass to carry it. If hominins were not disarticulating the animals, they likely either consumed the meat on the spot or they arrived after much of the meat was already gone. She concluded that scavenging, rather than hunting, may have been a critical part of their ecological niche.

The Rise and Demise of the Scavenging Hypothesis

Shipman outlined the necessary traits for a scavenging way of life. Scavengers must be able to travel long distances efficiently to locate carcasses and must have appropriately refined senses to do so. Bipedalism is relatively efficient at a jogging speed, but not exceptional for either walking or running. As visually oriented animals, hominins would have had to rely on what they see (as opposed to smell) from a distance. Perhaps watching for vultures would have been a useful strategy. At the carcass, they would have had to compete directly with large predators, or, as jackals and vultures do, meekly wait their turn. They may have contented themselves with backup plant foods in the meantime, but when they finally got their opportunity, they would have had to satisfy themselves with potentially spoiled meat and other leftovers that the predators could not exploit. Perhaps here is where stone technology would come in handy, smashing bones for marrow or brains, or, as the interest in nonmeat-bearing bones dares to suggest, in recovering resources besides food. Perhaps hominins treasured skin or sinew, important raw materials for making tools by modern peoples.

If hominin ancestors were serious about scavenging, would there have been enough food waiting to be discovered? Shipman's hypothesis, enthusiastically taken up by Robert Blumenschine, has inspired fieldwork that provides a better understanding of savanna ecosystems, as well as predator and scavenger ecology. While hyenas are more likely to devour prey almost in its entirety, the cats—including lions, leopards, and probably Plio-Pleistocene saber-tooth species—typically leave behind soft tissues and bones that may have been important resources for hominins. This can be mapped by walking across the landscape and looking for dead animals.

To obtain a better understanding of the activities at FLK *Zinj*, Bunn, Potts, and Domínguez independently undertook more extensive examinations of the bones. Excavations at this locality were triggered by the discovery of the cranium of *A. boisei* in 1959. Three hundred square meters of surface were exposed by removing the overlying rock, and the layer was dated by a tuff to about 1.75 Mya. The site yielded 2500 stone artifacts and tens of thousands of bones and fragments. Bunn found that the bones are skewed in their distributions, with long bones of the limbs being overrepresented for larger animals relative to foot bones, but not for smaller sized animals. This suggests a focus on the meat-bearing part of the animal. Furthermore, there was extensive breakage and fragmentation of the shafts of the long bones. When carnivores have first access to prey, they create tooth marks on over 75 % of the shafts, but they are less likely to break them. Scavengers arriving after the limb has been defleshed are more likely to chew on the ends, where cartilage and other tissues might remain and where it is easier to bite. Hominins, on the other hand, break into the marrow-rich shafts with hammer stones. That approach leaves intact epiphyses and fragmented shafts, as observed. At Olduvai, only 10–14 % of the shafts showed tooth marks.

Contrary to Shipman's observation, Bunn observed cut marks concentrated near the joints, which need to be disarticulated for butchering, especially for larger antelope. For smaller antelope, the frequency shifts to a greater percentage on parts of the skeleton where there would be more meat. These differences are consistent with the notion that larger animals had to be disarticulated. Separating the segment of a limb may have facilitated processing the bones for marrow or made the meat easier to transport. From all the evidence, Bunn and Potts each concluded that at this site hominins were bringing meaty bones to a central locality and then systematically butchering them with the stone tools. Hominins had early access to the carcasses, regardless of the unknown causes of death, and nonhuman scavengers had secondary access to the bones most of the time. This is not consistent with a model in which hominins had access only to the largely defleshed leavings of other predators.

In a follow-up study, Shipman compared the locations of the marks on Olduvai bones with marks made by stone tools from a Neolithic archaeological site. The later people made a far greater proportion of marks near the joints or on nonmeat-bearing bones. Disarticulation was more important to them, probably because butchery was followed by cooking.

The Paleolithic observations come from one site and one accumulation of bones, FLK *Zinj*. It demonstrates how some hominins behaved some of the time—it does not show what they did not do. Scavenging was certainly within their repertoire of foraging behavior, but further lessons have come from studying the competition. Hyenas are now recognized as some of the more dangerous predators, whereas lions would gladly chase away lesser predators and usurp their kills. When hunters scavenge and scavengers hunt, neither can afford to be too choosy about the source of their meals. Some of the most important adaptations for carnivory, processing the carcass and digesting the meat, would be needed by both predators and scavengers. The distinction between the two is greatly blurred.

Bone Composition and Diet

Which hominins were consuming meat? During the periods examined at Olduvai and East Turkana, both australopithecines and *Homo* were present. It is not possible to assign tools or cut marks to any one species. Beginning in the 1990s, anthropologists turned to bone chemistry for more information about diet.

Strontium is an element that can be incorporated into bones in place of calcium. The amount of strontium in plants varies according to the soils in different localities, and then it is reduced as it passes along the food chain. Thus, all other things being equal, the level of strontium relative to calcium is greater in the bones of herbivores than of carnivores.

Nitrogen and carbon are basic elements of biological molecules, but the isotopes present can reveal something about the diet. Nitrogen in the soil has a higher proportion of ^{15}N, whereas nitrogen in the atmosphere is almost entirely ^{14}N. Thus, low ratios of ^{15}N to ^{14}N are indicative of a marine food chain as opposed to a terrestrial food chain.

Two stable isotopes of carbon are common in the atmosphere, ^{12}C and ^{13}C. As plants take up carbon dioxide during photosynthesis, there are two different chemical pathways that may be used. Grasses and other plants adapted to more arid habitats, as well as some sedges in wetlands, often follow what is called a C4 metabolic pathway that preferentially takes up a higher ratio of ^{13}C. Other plants use a C3 pathway that has a lower proportion of ^{13}C. The isotopes, once captured by the plants, remain unchanged as they pass through the food chain into herbivores and beyond. Higher proportions of ^{13}C are indicative of grazing animals and the predators that feed on them.

When the teeth of early hominins are examined for their composition, it is seen that both *Australopithecus* and *Homo* were omnivorous, occupying a trophic level higher than primary herbivores and participating in both C3 and C4 food chains. Moreover, there was substantial variability from one individual to another and even within a single tooth. Diet probably changed with seasonal availability of resources and changing habitats. Apparently hominins always have been opportunistic about what they ate.

These tools are systematic ways of investigating diet through examination of fossil remains. Other clues have been gathered by unique and inspired investigations. For example, genetic analysis of human tapeworms has demonstrated that they diverged from their nearest relatives somewhat over 1 Ma ago. The life cycle of the tapeworm *Taenia* requires them to pass through two hosts, a carnivore and its herbivore prey, but species of worms are specific to particular hosts. It once was supposed that humans acquired the worms from their current hosts, domesticated pigs and cattle, sometime after the advent of agriculture. Instead it turns out that our parasites are more closely related to those that infest African cats and antelope. Thus, we seem to have acquired that particular affliction by sharing carcasses with lions.

Other dietary clues come from diverse sources. The earliest bone tools, from Swartkrans Cave, show wear suggestive of being used to excavate termite mounds. The fact that humans, like cats, are not able to synthesize the amino acid taurine well suggests a long evolutionary history of eating meat, which offers a ready supply of this essential nutrient. Likewise, the inability to manufacture vitamin C indicates a past diet in which this was not a necessity. A skeleton of *Homo ergaster* shows pathologies indicative of hypervitaminosis A. A lethal excess of this vitamin is most easily obtained by consuming the liver of carnivores, or possibly bee larvae, which concentrate large quantities.

Human are and long have been omnivores, although we now know this is an unwieldy label. Scarcely 50 years ago it was assumed that hunting played a crucial role in human evolution. As researchers began to investigate more closely the question changed repeatedly: Did early hominins hunt? Did they eat meat at all? Did they participate in the food chain of grassland or more wooded habitat? These questions have been answered only in general terms, and one could argue we still cannot describe early human diet very well. Nonetheless, the questions have steadily gotten more specific and have attracted many different disciplines to lend their tools to address them. Although we cannot point to a specific diet that helps to explain human evolution, we can now say with more certainty that there was no such single food type responsible. Instead, our ecological breadth itself is emerging as the defining evolutionary strategy.

Questions for Discussion

Q1: Why should anthropologists argue so much about hunting practices of ancient humans? What deeper preconceptions or values might be at stake?

Q2: Shipman suggested traits a species would need to have in order to be an effective scavenger. How are they different from the traits needed to be a successful hunter? Would these traits be less valuable or equally useful in species that do not eat meat?

Q3: Do humans show the traits of a scavenger or a hunter? Can they be found in fossil hominins?

Q4: How would nonhuman predators and hominins (or recent humans) have processed a carcass differently? Aside from cut marks, how could you tell whether a collection of fossil bones represented the leavings of people or of carnivores?

Q5: Dietary categories such as carnivore, omnivore, and frugivore are very simplistic. How would you describe the chimpanzee diet in a way that recognizes the differences from other frugivorous or omnivorous species, including humans?

Q6: Cultural diversity has complicated the question even more. What is the diet of the human species?

Q7: In light of such complications, what can we realistically hope to know about early hominin diet? What sort of answer would shed light on our evolution and adaptation?

Additional Reading

Blumenschine RJ, Cavallo JA (1992) Scavenging and human evolution. Sci Am 267(4):90–96

Bunn HT (1981) Archaeological evidence for meat-eating by Plio-Pleistocene hominids from Koobi Fora and Olduvai Gorge. Nature 291:574–577

Bunn HT (1986) Systematic butchery by Plio/Pleistocene hominids at Olduvai Gorge, Tanzania. Curr Anthropol 27(5):431–452

Domínguez-Rodrigo M et al (2007) Deconstructing Olduvai: a Taphonomic study of the Bed I sites. Springer, Dordrecht

Potts R (1988) Early hominid activities at Olduvai. Aldine de Gruyter, New York

Potts R, Shipman P (1981) Cutmarks made by stone tools on bones from Olduvai Gorge, Tanzania. Nature 291:577–580

Preutz JD, Bertolani P (2007) Savanna chimpanzees, *Pan troglodytes verus*, hunt with tools. Curr Biol 17:412–417

Shipman P (1986) Scavenging or hunting in early hominids: theoretical framework and tests. Am Anthropol 88:27–43

Stanford C (1999) The hunting apes: meat-eating and the origins of human behavior. Princeton University Press, Princeton

Case Study 13. Climate Change in the Pliocene: Environment and Human Origins

Abstract After our lineage diverged from those of the living apes, our ancestors ceased climbing in trees and began to walk bipedally on the ground. These behavioral and anatomical changes corresponded to shift in habitat, from arboreal to terrestrial; and this in turn appeared to correspond with more profound changes in the environment. A connection between climate and environmental change on the one hand and human emergence on the other hand has long been discussed, and the physical sciences have gradually improved anthropologists' ability to reconstruct past conditions with precision. It is now possible to track environmental fluctuations closely and attempt to map evolutionary change onto them. As data become more detailed, scientists continue to refine their questions. However, the causal relationship between climate change and human evolution is ultimately out of reach.

Beasts are of the forest, and human beings belong in cultivated clearings. These were the values of Europeans and colonial Americans who feared the dangers of the wilderness and who felt it a duty to God and civilization to tame it. Such pre-Darwinian prejudices served equally well to set forest-dwelling apes apart from their human relatives, and these same prejudices played a role in the early versions of human evolution, which saw descent from the trees and emergence from the forest as first steps in our story. Darwin put it this way:

> As soon as some ancient member in the great series of the Primates came to be less arboreal owing to a change in its manner of procuring subsistence, or to some change in the surrounding conditions, its habitual manner of progression would have been modified… (1872:433–434).

It did not appear necessary to speculate why human ancestors left the forest; coming to the ground was sufficient to explain the advent of bipedalism, tool use, brain expansion, and the other changes that made us human. A half-century later, when Raymond Dart named that ancestor *Australopithecus*, the dry grasslands that surrounded the caves in South Africa became the setting for the drama (Fig. 1). This was the Savanna Hypothesis, reiterated by many other anthropologists in the following decades.

J.H. Langdon, *The Science of Human Evolution*,
DOI 10.1007/978-3-319-41585-7_13

a

b

Fig. 1 (**a**) The savanna near Sterkfontein Cave and Johannesburg helped inspired Dart's version of the Savanna Hypothesis. (**b**) The savanna ecosystem contains a wide variety of subhabitats. This scene outside of Nairobi, Kenya, includes woodland and a watercourse amid grassy plains

Attention later focused on the Miocene Epoch as the time for human–ape divergence, and accumulating geological evidence made changing environmental conditions an important theme. Evidence from the ocean floor indicated that the second half of the Miocene and following Pliocene were periods of gradual change to cooler and drier climates, leading a million years ago to the Ice Ages in the northern continents. These studies have enabled us to pursue the association between a changing environment and our own history.

Tracking Past Climate Change

The modern standard for global temperature changes comes from studying oxygen isotopes in the shells of microscopic marine plankton called foraminifera. Two stable isotopes of oxygen are common in the water, the more common ^{16}O and a heavier ^{18}O. Chemically, they behave the same. However, the lighter ^{16}O evaporates more readily, so rain and freshwater have different proportions of the two isotopes than the ocean. When the earth passed through the Ice Ages, so much of the world's water became locked up in freshwater glaciers that the ratio of ^{18}O to ^{16}O in the oceans increased.

The isotope ratios in the oceans are captured by foraminifera when they construct their shells. As the organisms die, their shells create a perpetual and immense rain of sediment to the bottom. Those shells may be recovered in cores taken from the ocean floor to give us a record of isotope ratios stretching tens of millions of years into the past. We have found that there was a rapid increase in ^{18}O beginning in the Middle Miocene, about 15 Ma, and another increase starting about 5 Ma ago. These indicate the formation and expansion of ice sheets in the high latitudes, but those climate changes are also linked to Africa.

The deep-sea cores also contain deposits of dust that has blown off the continents. An increase in the amount of dust correlates with drier or drought conditions. Peter de Menocal reported that cores in the Atlantic and Indian oceans reflect conditions in West and East Africa, respectively. The magnitude of these deposits changes in regular cycles. Before 2.8 Ma, each cycle lasted 23–19 thousand years. After 2.8 Ma, the cycle shifts to a frequency of 41 thousand years. These cycles correspond to slight shifts in the earth's orbit that affect distribution of sunlight, global temperatures, and glacial expansion. The dust cores show that the cycles may also affect rainfall in tropical regions. The change in cycle frequency at 2.8 Ma apparently signals that expanding ice sheets in other parts of the world had reached a critical size so that they affected climate globally. The dust deposits also reach a peak in thickness about 2.8 Ma off the coast of West Africa. Another peak, in the Indian Ocean, comes about 1.7 Ma. A third peak, in both areas, occurs about 1.0 Ma. The sea core evidence shows us that the deteriorating climate in Africa involved drying as well as cooling.

Another signature of cooling is the presence of grasslands. We can track them through ratios of stable isotopes of carbon in the ancient soils (called paleosols).

Two stable isotopes are common in the atmosphere, ^{12}C and ^{13}C. As plants take up carbon dioxide during photosynthesis, there are two different chemical pathways that may be used. Grasses and other plants adapted to more arid habitats often follow what is called a C4 metabolic pathway that preferentially takes up a higher ratio of ^{13}C. Other plants use a C3 pathway that has a lower proportion of ^{13}C. The isotopes, once captured by the plants, remain unchanged as they pass through the food chain into herbivores and beyond. Isotope ratios may be measured for past environment by examining paleosols and also the fossil teeth and bone of animals in the food chain.

Thule Cerling and colleagues documented a global change in isotope ratios in mammalian teeth occurring between 8 and 6 Ma ago. This represents the retreat of forests and expansion of grasses and grazers in response to the Late Miocene cooling trend. It is a suggestive date, because it is roughly the time hominins diverged from the ape lineages. In East Africa, they report paleosol data for the widespread presence of C4 vegetation for the past 6 Ma.

East Side Story

African climate was certainly a part of the global pattern, but there were changes in Africa itself. While the world was cooling in a literal sense, the African Rift Valley was heating up tectonically. During the Late Miocene, motion between the eastern and western plates that make up the continent raised a double ridge of mountains with the Rift Valley between them. Active volcanoes added to these ranges. The effect was to isolate populations of animals on the two sides of the continent while creating changes in climate and vegetation.

Today the prevailing equatorial winds in Africa bring moisture from the Atlantic Ocean eastward across the continent. As those hot moisture-laden winds reach the mountains along the rift, they rise, cool, and drop their rain on the west. This creates the rain forests and feeds the Congo and related rivers. It also creates a rain shadow in the east. East Africa therefore is significantly drier, and grasslands flourish in place of rain forests.

In this context, French paleontologist Yves Coppens proposed the "East Side Story." Before the Middle Miocene, he argued, the African rainforest stretched continuously from coast to coast. As the mountains rose, they divided many species of animals, including human ancestors. In the west, where the forest continued as it was, the descendants of this ancestor did not need to change very much and became modern chimpanzees. In the eastern savanna, however, they had to invent a completely different suite of adaptations. In short, they became humans. The East Side Story was a restatement of the Savanna Hypothesis on a continental scale, using new knowledge of tectonic activity rather than global cooling to explain the same phenomenon.

Challenges to the Savanna Hypothesis

Even as climate data were being assembled to flesh out the Savanna Hypothesis, the model was running into trouble. New discoveries of fossil hominins showed that they were not living in the savanna. In the Afar region, where Lucy was found, fossils of mammals and shells indicated that it was a lake region with winding rivers and tropical forests. Other areas, such as the Omo Basin in Ethiopia and Kenya, were similar 3 Ma ago. The terrestrial animals and plants present, identified through bones and fossil pollen, indicate a variety of habitats, including closed and open woodland as well as grassland. Rivers and water sources in the savanna today often support narrow belts of trees—"gallery forests"—along their banks. Given the proximity of many different microenvironments, it is unclear which of these habitats Lucy preferred.

In 1994, two "new" older ancestors were described and named, pushing the record of human ancestry back another million years. *Australopithecus anamensis*, the probable ancestor of *A. afarensis*, lived along a river and lake system about 4.1–3.9 Ma. Fossils of this species from two sites in Kenya are accompanied by many aquatic species, including fish, crocodiles, and hippopotamus. However, as at Hadar and later sites, there are some animals present that prefer open country. Paleosols include carbon isotope ratios of plants more typical of semiarid or seasonal habitats.

Ardipithecus ramidus came from Aramis, Ethiopia, about 4.4 Ma in a more specific context. Among the species that accompanied it, aquatic animals were rare. The mammals such as woodland antelope and monkeys, along with pollen, fossilized wood, and sediments at the site indicate a forest setting.

Two additional species were named in 2002 that pushed known relatives back to 6 Ma. *Orrorin tugenensis* came from the Tugen Hills in Kenya. The fossils, which had been deposited in lake and channel sediments, came from open woodland with tree stands supporting smaller primates. *Sahelanthropus tchadensis*, from Toros-Menalla in Chad, was found with fossils that indicate a gallery forest, plus both aquatic and savanna species. Both lake and desert were nearby.

According to the Savanna Hypothesis, the shift to grasslands should have been a critical moment at the start of the hominin lineage with intense selection for open country adaptations. Instead, however only patches of savanna existed near our ancestors for their first 3 Ma of existence. Hominin fossils appear more consistently in context with woodland animals, but also nearly as consistently with those from multiple habitats. The term "mosaic environment" occurs repeatedly in the literature, suggesting patchy areas offering many possibilities. It is therefore not possible to associate the spread of the savanna with bipedalism or the divergence of the hominin lineage.

The Climate Forcing Model for *Homo*

The geological epochs were defined originally in part by characteristic fossils. Many Miocene species went extinct in the Pliocene and new species appeared. Paleontologist Elizabeth Vrba documented the turnover of fauna in South Africa and identified apparent waves of replacement of species. If the environment did change dramatically, one would expect some species to disappear and others to appear to take advantage of new opportunities. Rapid dramatic climate change should be indicated by major turnover events, and this would apply to hominins as well as to antelope. In Vrba's model, climate change forced the evolution of animal species.

Although her Climate Forcing Model has similarities to the Savanna Hypothesis, the timing is different. Vrba's pulse occurred about 2.5 Ma, corresponding to the start of the swing in oxygen isotope ratios and much later than the origin of hominins, but about right for the origin of robust australopithecines and *Homo*. In fact, that date appears to be crucial for a number of reasons. Between 3.0 and 2.0 Ma, in addition to the first *Homo* and *Paranthropus*, the beginnings of the Oldowan Culture and clear evidence of butchery of animals appear for the first time. At the end of that period, the brain is expanding in *Homo*, a number of human species are appearing, and humans are leaving Africa. In order to explore whether and how these events are linked with one another and with environmental change, it is necessary to explore the dating of all the events more precisely. Vrba's hypothesis focused attention specifically on the environment indicated by other types of mammals.

While Vrba's own data from South and East Africa supported the idea of a sudden replacement of fauna, other studies produced a less clear picture. Kay Behrensmeyer and colleagues did indeed confirm a turnover of about 50–60 % of species. However, their fine-grained analysis of East African sites produced mixed results. Some studies observe significant turnover events and others perceive gradual introduction of and elimination of species. Clearly change was occurring, but the pattern was complex.

Bovids are particularly useful for studying habitat change. Because they are relatively large animals, their bones fossilize well. Moreover, they are numerous and diverse, dominating Africa ecosystems of many types. It is possible to correlate species with individual habitats on the basis of skeletal morphology or with diets on the basis of jaws and teeth. For example, Lillian Spencer analyzed bovid jaw structure for adaptations for diet. She found it possible to distinguish between two types of grassland. Edaphic grasslands occur in areas of seasonal flooding, where periodic inundations favor dominance of plants that can recover quickly. Secondary savanna is the drier grassland usually considered in the evolutionary scenarios. Her bovid data suggested that edaphic grasslands probably had been typical of parts of East Africa for a long time, but that antelope adapted for secondary grasslands appear only about 2.0 Ma.

On a broader scale, ecological profiles can be established by looking at mammalian communities. Kaye Reed examined the composition of communities, including the percentage of species adapted for specific habitats or diets. Using modern ecosystems as a guide, she examined 27 Plio-Pleistocene fossil assemblages.

The transition from closed to predominantly open habitats was gradual over the period between 3.0 and 2.0 Ma ago.

When all of these data are assembled, it is clear that the East African environment fluctuated between greater and lesser rainfall and between more closed and more open habitats during the Pliocene and Pleistocene, even as there was an overall trend toward a drier ecology with a high proportion of grasslands (Table 1). Conspicuous climate shifts can also be tracked in local basins every couple of hundred thousand years. Changes even on this scale of appear to have a measurable impact on the mammals present.

How should one expect an ecological community to respond to climate change? A change in temperature or rainfall invites a new assemblage of plants to invade at the expense of those less tolerant of the altered conditions. Some herbivores might be able to change their diet but others would become more scarce or disappear from that region. Studies of carbon isotopes in teeth identified one bovid and one suid (member of the pig family) shifting to a diet of more C4 plants about 2.8 Ma, a time when the rest of the mammalian community was changing. Carnivores may be less affected by a change in herbivore prey. Thus, mammalian species may disappear from local habitats very quickly, but they could be thriving elsewhere. Local extinctions would be common, but species extinctions less so. New species may appear because they have migrated from neighboring regions, but evolution of new species would take time.

Finer grained studies continue to complicate the picture. Martin Trauth and colleagues collected data on lake levels in several basins in the Rift Valley between 2 and 3 Ma ago. Lake depths varied substantially and independently, affected not only by changing rainfall abundance but also by tectonic changes in the drainage areas. More importantly, these lakes changed independently of one another so that each basin had its own habitat history. At Lake Olduvai, where Olduvai Gorge now exists, Magill and coworkers documented for distinct cycles of increasing C4 vegetation in the short interval between 1.9 and 1.8 Ma, from which a number of fossil hominins came.

As a further complication, a fossil assemblage may represent bones accumulated over a few thousand years, perhaps longer. The fossils thus represent all the habitats in that locality through a period of time rather than a single habitat or a picture frozen in time. This phenomenon is known as time averaging, and in a rapidly fluctuating environment, this may give the appearance of a mosaic habitat. It makes it difficult for scientists to associate a specific rare species, such as a hominin, with a particular habitat in a single example.

By combining data from multiple sites where hominins exist, it may be possible to gain a clearer understanding. *A. afarensis* is known from different sites in Ethiopia, Kenya, and Tanzania over a period of six hundred thousand years, from 3.6 to 3.0 Ma. During that interval, the environment, as revealed by pollen and other studies, changed several times. *A. afarensis* may have been a highly tolerant species. Environmental changes after 3.0 Ma apparently were more drastic or crossed some critical threshold of rainfall, because there were greater changes in mammalian species. By examining community structures associated with hominins, Reed identified some preferences. The earlier australopithecines, *A. afarensis* and *A. africanus*, lived in well-watered woodlands. Robust australopithecines were

Table 1 Ecological and evolutionary changes in East Africa 3.0–1.7 Ma

Date	Paleoenvironment wetter	Paleoenvironment drier	Evidence	Hominin evolution, first appearance
3.5–3.3 Ma		Increasing aridity	Carbon isotopes of paleosols	*Australopithecus afarensis* present
		High diversity of bovid species		
3.3–3.0 Ma	Higher rainfall, cooler		Pollen	*Australopithecus afarensis* present
	High diversity of bovid species		Carbon isotopes of paleosols	Oldest known tools (3.3 Ma)
			Bovid species	
2.8 Ma		Increasing aridity	Oceanic dust deposits (West Africa)	
		Increase in open habitats	Bovid species	
		Bovid species turnover event	Carbon isotopes in mammalian teeth	
2.7–2.5 Ma	Increased rainfall		Expansion of lake sediments	Stone tools (2.6 Ma)
	High diversity of bovid species		Bovid fossils	Earliest Homo (2.7 Ma)
2.5–2.3 Ma		Increasing aridity	Carbon isotopes of paleosols	*Paranthropus aethiopicus* (2.5 Ma)
		Heterogeneous habitats	Bovid species	*A. garhi* (2.5 Ma)
		Opening of savanna	Faunal community structure	*Homo* sp. indet.
				Evidence of carcass processing
2.3–2.1 Ma	High habitat diversity		Bovid species	*Paranthropus boisei* (2.3 Ma and after)
				H. rudolfensis (?2.4–1.9 Ma)
				Brain expansion
1.9–1.7 Ma	Increased rainfall	Heterogeneous habitats	Expansion of lake sediments	*H. habilis* (about 2.0 Ma and after)
	High bovid diversity	Appearance of open habitats	Carbon isotopes of paleosols	*H. ergaster* (about 1.8 Ma and after)
			Carbon isotopes of lake sediments	Modern limb proportions
			Bovid fossils	
1.7 Ma	Increasing aridity		Carbon isotopes	
	Faunal turnover		Fauna at Olduvai	

found in these and also more open habitats, especially grasslands that flooded seasonally. When *Homo* was present, the environment was more likely to include open, drier grasslands, but commonly near lakes or rivers. Stone tools, which are not dependent on wetlands for preservation, have been reported from both wooded and grassland settings. These studies suggest interesting scenarios of ecological adaptation, but do not clarify a specific role for the environment driving hominin evolution.

Variability Selection

Richard Potts took a different perspective. While some studies were trying to link specific environmental events with species evolution, Potts was more impressed by the increasing instability of the environment in the past 5 Ma. A species that adapted specifically to a new, drier habitat would not survive long, as that environment would be prone to change again in a geologically short period of time. What might be a more important adaptation is the ability to thrive in a wide range of habitats and conditions. Such a strategy produces an ecological generalist.

While we think of a generalist as a species lacking specialized adaptations, Potts envisioned selection for traits that supported the ecological plasticity needed to tolerate an unstable and changing environment. He called this variability selection. His list of adaptive characteristics reflect flexibility of behavior—adaptability of locomotor systems, diet, foraging strategies, information processing, and social structures. Unfortunately, it remains impossible to investigate a precise link between climate and the origin of *Homo*. We have only a handful of fossils for the first half million years of our genus, too few to pin down the time and place for its origin and too incomplete to assign to species. The variability selection model appears to fit our own genus well, but it may also reflect the absence of detail in our knowledge of hominins during this crucial stage.

Conclusion: Finding the Right Questions

A century of data collection has provided a much greater understanding of past environments and climate changes. Although many questions have been answered, there are always more to address. In what setting did humans evolve? Diverse data show complex environments with many subhabitats. Hominins probably exploited many of them. Did the environment change in East Africa? Animal fossils, pollen, and soil isotopes confirm that it did, but in complex ways with an extended period of unstable transition. How close was the link between climate change and faunal change? There seems to be a reasonably close link for bovid communities, but the number of fossil hominins is too small to address that question for them. The earliest hominins probably did not evolve on the savanna, but open woodlands and grasslands became important for *Homo*.

The questions we can ask and answer about paleoenvironments grow increasingly more detailed. The most important question, however—did climate change inspire human evolution?—is a different type of question. It asks why rather than what and is probably not answerable by science.

Questions for Discussion

Q1: Rainfall patterns can change quickly over a period of decades. If we could track fluctuations at that level of resolution through lake sediments, could we improve our understanding of evolution?
Q2: *Australopithecus afarensis* persisted through hundreds of thousands of years of climatic instability. Is this evidence for or against climate forcing?
Q3: Why can't science give us a definitive answer to the question of whether climate change caused human evolution?
Q4: Will global warming in the present and near future have an evolutionary impact on modern humans?
Q5: If we are an adaptable species and have the potential for further evolution, why should we worry about climate change today?

Additional Reading

Behrensmeyer AK (2006) Climate change and human evolution. Science 311:476–478
Behrensmeyer AK et al (1997) Late Pliocene faunal turnover in the Turkana Basin, Kenya and Ethopia. Science 278:1589–1594
Bibi F et al (2013) Ecological change in the Lower Omo Valley around 2.8 Ma. Biol Lett 9(1):20120890
Bobe R, Eck G (2001) Responses of African bovids to Pliocene climatic change. Paleobiology 27(sp2):1–48
Bobe R et al (eds) (2007) Hominin environments in the East African Pliocene. Springer, Dordrecht
Cerling TE et al (2011) Woody cover and hominin environments in the past 6 million years. Nature 476:51–56
Coppens Y (1994) East side story: the origin of humankind. Sci Am 270:88–95
Darwin C (1872) The origin of species and the descent of man. Modern Library, New York
de Menocal PB (1995) Plio-Pleistocene African climate. Science 270:53–59
Johanson D, Edey M (1981) Lucy: the beginnings of humankind. Simon & Schuster, New York
Magill CR et al (2013) Ecosystem variability and early human habitats in eastern Africa. Proc Natl Acad Sci U S A 110:1167–1174
Plummer TW et al (2009) Oldest evidence of toolmaking hominins in a grassland-dominated ecosystem. PLoS One 9, e7199
Potts R (1996) Humanity's descent: the consequences of ecological instability. AvonBooks, New York
Potts R (1998) Variability selection in hominid evolution. Evol Anthropol 7(3):81–96
Quinn RL et al (2013) Pedogenic carbonate stable isotope evidence for wooded habitat preference of early Pleistocene tool makers in the Turkana Basin. J Hum Evol 65:65–78
Reed K (1997) Early hominid evolution and ecological change through the African Plio-Pleistocene. J Hum Evol 32:289–322
Shreeve J (1996) Sunset on the savanna. Discover 17:116–125
Spencer LM (1997) Dietary adaptations of Plio-Pleistocene bovidae: implications for hominid habitat use. J Hum Evol 32:201–228
Trauth MH et al (2005) Late Cenozoic moisture history of East Africa. Science 309:2051–2053
Vrba ES (1993) The pulse that produced us. Nat Hist 102(5):47–51
Vrba ES et al (eds) (1995) Paleoclimate and evolution with an emphasis on human origins. Yale University Press, New Haven

Case Study 14. Free Range *Homo*: Modernizing the Body at Dmanisi

Abstract The differences between australopithecines and *Homo* are more than brains and teeth—compared with living apes, the rest of the body underwent a transformation as well. Skeletally, humans have different body proportions even from the australopithecines. Soft-tissue organs also have unique characteristics. Although the fossil record does not provide direct information about those, we can look for functional patterns that might relate them to skeletal changes. The earliest fossils of *Homo* outside of East Africa suggest a reorganization of body design that made us world travelers.

The continued evolution of the body below the neck deserves some investigation. Modern humans do not walk as australopithecines did, and they have steadily departed from the adaptations that made our ancestors adept at living in trees. What did we gain? Longer lower limbs, less mobile joints at the ankle and within the foot, a rigidly adducted first toe, and other changes argue for more efficient walking and running. Efficiency becomes more critical as the amount of time and effort invested in walking and running increases. Thus, several authors have interpreted these adaptations in terms of long-distance travel or endurance running, but it is not the bones alone that explain this adaptation.

Breathing and Thermoregulation for Endurance

Humans are not the fastest species, but people in good physical shape can match or outperform other mammals in the duration of time and distance. David Carrier was one of the first to look systematically at the evolutionary importance of endurance running. Studies of respiration he had conducted with physiologist Dennis Bramble showed that bipedal locomotion is free of constraints on breathing that quadrupedal animals experience. Most mammals involve trunk flexion and diaphragm contractions in their different gaits. These actions force expiration and inspiration in rhythm with stride so that oxygen availability is more a reflection of speed and lung

J.H. Langdon, *The Science of Human Evolution*,
DOI 10.1007/978-3-319-41585-7_14

properties than of metabolic needs. Consequently, most quadrupeds are incapable of making fine adjustments of breathing depth and rate.

Typical quadrupeds, such as horses, have a preferred speed for a given gait, whether walking, trotting, or galloping. As the animal speeds up or slows down from the optimal speed, it must change gait at certain thresholds. When forced to move at rates in between these preferred speeds, efficiency (measured in oxygen consumed per distance traveled) goes down and the metabolic cost increases. Inefficient locomotion consumes oxygen faster than it can be supplied by the lungs, thus requiring the body to undergo anaerobic metabolism. Generating energy without an adequate flow of oxygen builds up an oxygen debt and accumulates waste products that detract from performance. In contrast, humans maintain the same efficiency across a wide range of running speeds. Because our breathing is controlled by the diaphragm independently of locomotion, we can adjust our rate and depth of breathing to better match actual oxygen demand by body tissues and run for hours at a time. Trained athletes can maintain aerobic running at speeds comparable to preferred speeds of many larger animals. Especially when animals are forced to move at non-preferred speeds, humans in good condition have far greater endurance.

Endurance is also affected by the ability to regulate body temperature, as exercise generates heat. Human skin has a far greater capacity to dissipate extra heat than that of most other mammals. With most of the hair eliminated and an increase in the number and distribution of sweat glands, water may be secreted onto the surface of the skin where it can absorb heat as it evaporates. Several specific adaptations work together for this. Excess body heat can be radiated from the surface of the skin into the air. Likewise, radiant heat from the sun or terrestrial environment can be absorbed by the skin. Fur forms an insulating layer that prevents air from circulating close to the skin of most mammals and blocks radiation of heat in both directions. It thus helps to maintain a constant temperature despite fluctuations in the external environment; however, it does not respond to changes in the internal environment. By eliminating fur, our own bodies increase exposure to the hot sun and chill of the night, but also creates tolerance of the body's internal states.

Human skin has the unique ability to direct greater or lesser amounts of blood flow to the surface. Constriction of arterioles in the dermis keeps the most of the blood deep to a layer of subcutaneous fat so that heat is retained. The distribution of fat itself is unusual among mammals. Although we concentrate superficial fat in the same deposits as other mammals, those deposits are more extensive and underlie a far greater proportion of the skin than in other species. They provide some insulation to conserve heat in deeper tissues. In order to dump excess heat, the arterioles are opened to that considerably more blood flows to the surface and heat is radiated away. This mechanism produces a visible reddening of pale skin—thus Mark Twain's famous quip, "Man is the only animal that blushes, or needs to."

A second cooling mechanism is perspiration. Human sweat is produced by eccrine glands that are spread liberally across the body. These are more restricted in most animals to hairless areas on the feet and around the nose, probably because moisture secreted under fur would not evaporate easily. Evaporative cooling supplements radiation of heat, but does so at the cost of losing water, salt, and other electrolytes. Some animals, including dogs, dump excess heat through the mouth by panting, which facilitates the evaporation of saliva. However, panting interferes with the deeper breathing needed to supply more oxygen, and it is incompatible with respiration driven by body actions during running. Therefore, panting has a limited ability to cool an animal still exercising.

Eccrine glands in other mammals are stimulated in response to emotional state. The glands in our own hands and feet still respond in this way; but those in the rest of the body secrete in response to rising skin temperature. Thus, they help the body and individual parts of the body prevent the accumulation of extra heat.

Eccrine glands work like individual nephrons, filtrating blood to produce perspiration. Although they are less capable of recovering salts and other useful molecules from the blood, they may provide a supplementary mechanism to eliminate metabolic waste.

Finally, we have a complex pattern of venous circulation around the brain that mixes with surface blood from the face and scalp to help cool the brain. The brain is extremely sensitive to changes in temperature. Small increases may cause it to cease functioning, as may occur in heat stroke or fevers. We are not the only species with circulatory adaptations to cool the brain, nor the most efficient at it; but such strategies are necessary to increase the tolerance of rising body temperature during exercise. Brain temperature may be a critical limiting factor for endurance.

The human capacity to cool our bodies as we exercise increases our endurance, but it is also costly. We need to replenish fluids on a regular basis and are more dependent on living near water sources than are other animals. Without insulating fur, we depend more on metabolic activity to maintain body temperature when resting or when the air is cool. This demands an increased supply of calories, just as does the exercise itself. We therefore require a richer diet, more dense in calories. One of the other unusual characteristics of human skin is the body-wide distribution of subcutaneous fat reserves. It is difficult, and probably meaningless, to attempt to determine a "normal" amount of body fat for the human species, given the extreme variance of diets, activity levels, and physique among different cultures. While it may assist in conserving heat, variations in thickness and unprotected area of the body argue that this is unlikely to be its primary function. What is important is the ability of individuals, especially infants, to store energy relatively easily for later use. For infants, fat is the critical buffer that allows the brain to continue its development even when food supplies may be inconsistently available. For the rest of us, energy stores help sustain high expenditures of energy over long periods of time.

A Skeleton for Endurance

Daniel Lieberman and Dennis Bramble identified a number of adaptations unique to the redesign of the human musculoskeletal system that favor endurance running. Lower, laterally directed shoulders are more consistent with the counter-rotation of the upper body compared to the lower parts. The counter-rotation stores energy within the body and increases efficiency of both walking and running. Shorter, lighter upper limbs save energy during running, while longer lower limbs increase stride length, speed, and running efficiency. The expanded surface areas of lower limb joints and the increased volume of the calcaneus absorb and distribute stresses.

In the musculature, the long Achilles tendon and structures supporting the longitudinal arch serve as energy storage devices that absorb shock and also increase efficiency. Gluteus maximus, one of the more distinctive muscles in humans, has been reoriented and strengthened for powered hip extension. It is usually relatively inactive during walking, but becomes important when running or climbing.

A taller, more linear body shape facilitates the dispersal of body heat across a proportionately greater surface area. Enlarged semicircular canals help stabilize the head and body and appear to develop along with the unique challenges to balance that bipedalism create. At the least, these traits describe how we have achieved efficient bipedal locomotion. Very likely, according to Lieberman, they represent selection for endurance running as well.

Endurance and Human Evolution

Why was endurance so important? Two answers have been commonly cited: foraging strategies involving running or long-distance travel. Carrier argued for endurance hunting—the human ability to chase animals until they collapsed from exhaustion. This might have been one strategy for bringing down game before the invention of effective throwing weapons. Although there are anecdotes of endurance hunting by native peoples in different parts of the world, it is a relatively rare endeavor today. Pat Shipman suggested the importance for scavenging hominins of being able to run quickly to an animal carcass in order to compete with other scavengers. As discussed in Case Study 12, however, scavenging was unlikely to have been a primary adaptation for us.

Foraging by modern hunter-gatherers requires walking long distances on a daily basis, in search of plant foods, firewood, and workable stone as well as game. Many factors combine to increase those distances. Feeding a band of large mammals requires sizable home range of resources to exploit. An active life style and a large brain demand an even greater intake of quality food items, including meat when possible. Tolerance of distance enabled hominins to cross undesirable habitats in search of better ones; thus, an important consequence of travel was the ability to expand the species range. The African fossil record is not complete enough for us to be able to follow hominins across the landscape, but the appearance of humans in Asia indicates a critical stage in the development of our ecological position in the world. If both skeletal and soft tissues traits described earlier are correctly interpreted, they may well have evolved at about the same time. Some of those changes are apparent in the oldest human fossils known from Asia with evidence of human presence, at Dmanisi, in the Republic of Georgia.

Dmanisi

Dmanisi was a medieval city in the Caucasus Mountains astride the Silk Road, from which it drew its wealth. Twentieth-century archaeologists have long been interested in the historic ruins, but while they were excavating grain storage pits they stumbled upon bones of early Pleistocene mammals. As the focus of the excavations shifted to paleontology, anthropologists led by David Lordkipanidze unearthed primitive stone tools and, in 1991, a human mandible. The jaw, dated to 1.78 Ma, was followed by the discovery of five crania and other bones. Along with these are thousands of bones of other animals and more than a thousand stone tools comparable to the Oldowan tradition in Africa. The fossils have been securely dated by a variety of means, including radiometric dating of underlying and overlying layers of volcanic basalt and a paleomagnetic reversal correlated with the Olduvai Gorge sequence. The dates are consistent with the fauna present. The period between 2.3 and 1.7 Ma was a relatively warm and wet phase and may have facilitated the expansion of the hominin range. Dmanisi at the time is reconstructed to have mixed woodland habitats ranging from gallery forests to savanna, based on the animals present.

Identifying the species of *Homo* is problematic. In East Africa at this time, three species are recognized, although sorting bones among them is difficult. Of these, *H. ergaster* is most derived and is believed to be ancestral to later *Homo*. It is considered by some to be synonymous with later *H. erectus* known from Asia and Africa. By this logic, the hominin at Dmanisi should be *H. ergaster* or *H. erectus* or some transitional form. However, the situation is more complicated. It would appear improbable that this one site in Georgia contains more than one species, yet the known specimens show great variability. For example, the five measurable crania have endocranial volumes (braincase volumes) of 546, 601, 641, <650, and 730 cm^3. The smaller four overlap with *H. habilis*, but fall below known specimens of *H. rudolfensis* and *H. ergaster*, while the fifth is comparable to *H. ergaster* but exceeds known *H. habilis. H. ergaster* and *H. erectus* are distinguished by a number of cranial features. *H. ergaster* has prominent brow ridges that take the form of two distinct arches above the eyes; *H. erectus* more commonly has a single, heavier brow that spans the width of the face. While some of the Dmanisi crania show a single brow ridge, one has only a small brow, more similar to *H. habilis*. Details of the first mandible resemble *H. ergaster*, but the sequential reduction of molars from front to back is a derived feature seen in *H. erectus*. Researchers at Dmanisi assigned the first discovered mandible to a new species, *Homo georgicus*; but putting the specimens in a separate species does not resolve relationships nor explain the diversity among them. These apparent inconsistencies make more sense if the emigrant ancestor had separated from the African lineages at a very early date. The Dmanisi specimens are roughly the same age as the earliest known African specimens of *Homo* that are complete enough to be identified to species. We can assume that they shared an older and probably more primitive common ancestor.

The skeleton below the neck is equally challenging. While there are striking contrasts to the body design of *Australopithecus* and later *Homo*, Dmanisi is perhaps the closest representative we have of a transitional form in the direction of modern proportions. There are a number of primitive features present, especially in the upper limb. The shoulder blade points more upward, like those of *Australopithecus* and the great apes, and the humerus lacks torsion of the shaft that characterizes later humans. However, the rest of the skeleton shows more modern features. The incomplete spine shows a hint of lumbar lordosis (curvature in the small of the back) that we associate with fully upright posture. Lower limb length and limb proportions (femur to tibia, humerus to femur ratios) are in the range of modern humans. Overall the individuals represented by partial skeletons are small. They are estimated from long bone lengths to have been between 145 and 166 cm and to have weighed 45–60 kg.

Australopithecus expressed bipedal locomotion primarily in the lower part of the body, with climbing adaptations retained in the relatively long and robust upper limbs. But its lower limbs were short and had many unique features. The Dmanisi hominins also show a mosaic evolution of the limbs. The lower limbs and feet are approaching modern morphology in most aspects, including their greater length, but the upper limbs lag behind. The more lateral orientation of the modern shoulder that facilitates throwing, for example, and other manual activities are not yet present.

The Dmanisi excavations are not complete and one can expect more material to come to light. The hominins there are a fascinating reminder that perhaps not all the important events of human evolution occurred in Africa. They provide a temporal context both for reorganizing the skeleton and for expansion across Eurasia. They may also help to place in time-related changes in body hair and soft tissue physiology and corresponding behaviors. A skeleton that is even more modern in its design is the remarkably complete adolescent *H. ergaster* from Nariokotome, Kenya, just over 1.5 Ma, discussed in the following chapter.

Questions for Discussion

Q1: What other animals have exceptional endurance? What ecological circumstances might explain this?

Q2: How could one distinguish between and test the two hypotheses for endurance, running vs. long distance travel?

Q3: How might the fact that hominins were in the Republic of Georgia by 1.78 Ma change our understanding of the role of East Africa in human evolution?

Q4: Several species of *Homo* appear to exist together in East Africa. Could this also be true in Georgia? What would be the implications of that for evolutionary and ecological relationships among them?

Q5: The Dmanisi crania show morphology that seems to combine traits of *H. habilis*, *H. ergaster*, and *H. erectus*. Assuming only one species is represented, what possible interpretations of phylogenies can account for this?

Additional Reading

Bramble DM, Lieberman DE (2004) Endurance running and the evolution of *Homo*. Nature 432:345–352

Carrier DR (1984) The energetic paradox of human running and hominid evolution. Curr Anthropol 25(4):483–495

Gabunia L et al (2001) Dmanisi and dispersal. Evol Anthropol 10(5):158–170

Langdon JH (2005) The human strategy: human anatomy in evolutionary perspective. Oxford University Press, New York

Lordkipandze D et al (2007) Postcranial evidence from early *Homo* from Dmanisi, Georgia. Nature 449:305–310

Wong K (2003) Stranger in a strange land. Sci Am. 74–81

Case Study 15. Reading the Bones (3): Tracking Life History at Nariokotome

Abstract In 1984 a team led by Richard Leakey and Alan Walker discovered a remarkably complete skeleton of *Homo ergaster* on the western side of Lake Turkana in Kenya. It was pieced together from small fragments that had been shattered by tree roots and scattered across the barren ground. When finally assembled, the bones were found to belong to a boy initially estimated to be about 12 years old. However, determining exactly how old he was at the time he died raises important questions about rates of maturation, brain development, and life history strategies in early hominins. One and a half million years ago it appears that humans had not yet acquired one very distinctive characteristic of modern populations—childhood.

The Nariokotome skeleton (KNM-ER 15000) was recovered among some of the most intensively examined hominin deposits in the world, near Lake Turkana in Kenya. On both east and west sides of the lake and to the north where the Omo River empties into it, hominin fossils and stone tools and bones of many other animals occur in abundance. This particular find occurred on the west side of the lake, between two securely dated volcanic tuffs from 1.88 to 1.32 My. Sediments here were deposited in a seasonal flood plain. By assuming a constant rate of accumulation (a convention known to be imprecise but reasonable when dates between strata are close together) the position of the fossil indicates a date between 1.56 and 1.51 My.

The Age of Nariokotome Boy

The Nariokotome boy is not the only immature individual known. The type specimen of *Australopithecus* is a juvenile from South Africa. Raymond Dart compared that to a 6-year-old human child. The first specimens of *H. habilis* from Olduvai come from an adolescent of 12 or 13 years by human standards. Neanderthal skeletons represent a range of ages from infants to adults. Physical anthropologists have long been studying developmental changes in the skeleton, but there are difficulties applying modern human standards to earlier species.

J.H. Langdon, *The Science of Human Evolution*,
DOI 10.1007/978-3-319-41585-7_15

Determining the developmental age of a skeleton depends on identifying changes that occur at predictable rates or ages. One of the most reliable sequences of changes for nonadult individuals involves the development and eruption of teeth. Crown formation, root formation, and eruption occur for each tooth in a regular pattern. All may be readily detected on X-ray images. Fortunately, all of the Nariokotome boy's teeth are present, except for the unformed third molars, or wisdom teeth. The upper permanent canines have not erupted and the deciduous canines are still in place. Most of the teeth, however, were not completely formed, as the roots were still growing. One of the upper third molars is visible on X-ray still within the bone. It is therefore possible to present an independent estimate of developmental age for each tooth, based on modern human standards. Those estimates will vary depending on the human population to which the specimen is compared. B. Holly Smith has assembled this data and evaluated the fossil. Part of her analysis is presented in Table 1 comparing the fossil to one of her comparison groups (North American white males) and also to great apes. Most of the teeth indicate a developmental age of 10–11 years. Using other reference populations or patterns of dental maturation does not alter the results substantially.

Table 1 Estimation of dental age of the Nariokotome fossil on the basis of human and great ape samples (Smith 1993)

Tooth	Development in KNM-ER 15000	Age on human scale (years)	Age on great ape scale
Maxilla			
I1	Root fully developed	At least 10.6	At least 6.5
I	Root length complete, apex not closed	10.1	6.2
C1	Root length two-thirds complete	9.5	8.2
P3	Root length two-thirds complete	9.9	6.6
P4	Root length three quarters complete	10.6	7.0
M1			
M2	Root length two-thirds complete	11.4	6.6
M3	Crown incomplete	12.3	6.7
Mandible			
I1	Root fully developed	At least 9.2	At least 6.5
I2	Root fully developed	At least 9.9	At least 6.7
C1	Root length three quarters complete	10.2	8.6
P3	Root length half to two-thirds complete	10.0	6.4
P4	Root length half to two-thirds complete	10.5	6.6
M1	Root fully developed	At least 10.0	At least 5.7
M2	Root length half complete	12.3	6.2
M3	Crown incomplete	10.7	6.7
Average dental age		10.7	6.9

Anthropologists must be cautious in evaluating the fossil by a modern human scale. Humans do mature more slowly than do other primates, so this is likely to produce an upper limit on age. In one obvious way, the Nariokotome boy differs from modern people. Normally the modern human canine erupts a year or so before the molar, but in the fossil, the upper second molar has already erupted, while the deciduous upper canine has not yet been lost. The fossil closely reflects the eruption sequence seen in apes, where the much larger canine teeth take longer to develop. Thus, the second molar comes in earlier (at least 7.5 years) and the canine much later. Using more comprehensive data from the teeth, Smith estimated the Nariokotome specimen to have a chimpanzee dental age of 6–7 years based on the molars and other teeth, but more than eight according to the canines. These figures define a likely lower age limit.

Another means of determining developmental progress is to examine the fusion of elements of the bones. Most of the bones of the body are first created from cartilage, which is a softer embryonic tissue more capable of growth. Within that cartilage, one or more centers of ossification will appear where the cartilage degenerates and is replaced by bone. The centers of ossification expand until they replace all the cartilage.

Long bones of the body typically have at least three centers of ossification. One begins in the middle of the shaft. Usually the joint surface on each end, called an epiphysis, ossifies separately so that the joint is supported by strong bone from an early age. As a child grows older, those ossification centers expand toward one another. The cartilage between them may continue to grow, adding length to the bone. Cartilage growth is stimulated by growth hormone, while ossification is accelerated by sex hormones. As the child enters adolescence, there is a surge of growth hormone, corresponding to a rapid increase in height. Puberty is caused by a greater release of sex hormones, and the spreading ossification centers begin to overtake the growing cartilage. Growth ceases when the ossification centers meet and the epiphyses fuse to the shafts of the bone. Body height stops increasing about age 18–20, earlier for girls than for boys. The timing of epiphyseal fusion will vary for different bones. Various factors can alter growth rates. For example, malnutrition may not allow tissues to respond vigorously to growth hormone, leading to smaller stature by the time growth ceases. Good nutrition may permit a person to achieve maximal growth. Hypernutrition, especially a steady surplus of calories, may cause rapid growth and tall stature, especially in childhood, but it may also facilitate an early puberty and thus an early cessation of growth. These factors introduce some uncertainty into aging a skeleton.

The Nariokotome skeleton has bones with centers of ossification in varying degrees of development. Each of these provides an independent comparison to the modern human pattern. For example, the three primary units of the coxal bone in the pelvis—the ilium, ischium, and pubis—have not yet fused. Fusion of the coxal bone normally begins around age 9–12 and is completed around age 14–18. Some of the elements of the humerus have fused; others had not. This would place the Nariokotome child between 12–15 and 14–16 years. Overall, using such information, Smith estimated a skeletal age for the fossil at about 13–13.5 years with some uncertainty. However, when chimpanzees were used as a reference, she calculated a skeletal age of about 7.5 years.

When the results of dental and skeletal studies are combined, there is some discrepancy between them on the human scale. The skeleton appears more advanced than the teeth. The chimpanzee scale appears to produce more consistent results comparing the overall dental and skeletal ages but runs into greater problems with internal correlations. The age estimates from the canines are greatly out of line with those from other teeth. This is to be expected because the canines of early *Homo* are already reduced in size and thus take less time to develop. We are left with the unsurprising conclusion that neither model fits the fossil perfectly. Instead, the fossil fits well as one point in an evolutionary spectrum that connects chimps and humans. More specifically, the developmental schedule fits between humans and what we know of *Australopithecus* development.

Pinning Down the Rate of Development

Understanding the absolute age of Nariokotome boy is particularly important because humans and chimpanzees develop at substantially different rates. A debate had already been raging over whether australopithecines showed a chimp-like rapid maturation or a human-like slow one. The Nariokotome skeleton provided fresh evidence on that question and on the related issue of when the evolutionary change occurred. To answer that question of maturation rate, one not only needs to know the developmental age of the fossil, but also its absolute chronological age at the time of death. The tools for determining this were developed in the decade after the discovery.

Dental enamel is laid down in a daily cycle during the period of tooth crown formation. Daily deposits can be observed under an electron microscope as striations on the enamel. It is possible to count them as one would count rings on a tree and thus to compare enamel formation times of living and fossil species. Hominins and apes differ in enamel thickness, but this difference is independent of the differences in rate. Apes deposit enamel faster and they complete crown formation more quickly.

Chris Dean and his colleagues applied this technique to examining fossil hominins. The australopithecine and early *Homo* specimens they examined laid down enamel as slowly as African apes, but took somewhat longer simply because the final enamel thickness was greater. For example, the Nariokotome specimen completed occlusal enamel deposition on the upper medial incisor in about 212 days, while modern humans take over 289 days, a third again longer. Other teeth show similar differences. Although the pattern of skeletal development for Nariokotome may resemble modern humans in many ways, the absolute rate was much faster. In chronological age, the fossil was probably 8 years old.

How fast an individual grows and how long he or she lives is the result of many evolutionary trade-offs. There are reasons to grow up quickly. The faster an individual matures, the sooner he or she can begin to reproduce, and there is definitely an advantage to getting a step ahead of the competition. The probability of dying before getting a chance to reproduce—because of disease, a fatal accident, or being eaten by a predator—increases the longer one puts off puberty. On the other hand, maturing slowly gives one time to grow bigger. Bigger may mean one is safer from predators or more likely to succeed in the all-important competition for a mate. Maturing slowly gives one more time to set aside energy and nutrient reserves to spend on the next generation.

Humans grow slowly. Our large brain is consistent with delayed maturation and a long life. We do have a few unique aspects, though. Our gestation period is less than we might predict from our brain size—9 months versus 18 months, according to one estimate. We wean our children early—as late as 5 years in some populations, but more typically 2 or 3 years in nonindustrialized populations. As a result, all humans all spend a period of their lives when they are not nursing, but are still totally dependent on adults for food and survival. Childhood defined in this way is unique among animals. It has profound implications for our social organization and economy, since parents have to invest in their children much longer and more than one adult is needed to provide for a child. However, it ultimately increases fertility. The mother potentially can start her next child sooner and have overlapping dependent children.

When did life history change to the modern pattern? If the Nariokotome boy matured at the rate indicated by his bones and teeth, he likely did not experience much of a childhood. Although the brain was getting larger in early *Homo*, the human pattern of slow development arrived much later.

Questions for Discussion

Q1: How does one tell that a mammalian skeleton comes form an immature animal?

Q2: The upper canine tooth was the last to be replaced in the Nariokotome boy and was the most troublesome in determining the dental age. What is unusual about the canine in human evolution that would explain this?

Q3: Why do species have so many different life history strategies? Why isn't there one best strategy?

Q4: What are the costs and benefits of having a large brain? Why do we have one? Why don't more species?

Q5: What are the costs and benefits of childhood from the perspective of the child? of the mother?

Additional Reading

Dean MD (2006) Tooth microstructure tracks the pace of human life-history evolution. Proc Biol Sci 273:2799–2808

Dean MC et al (2001) Growth processes in teeth distinguish modern humans from *Homo erectus* and earlier hominins. Nature 414:628–631

Smith BH (1993) The physiological age of KNM-WT 15000. In: Walker A, Leakey R (eds) The Nariokotome *Homo erectus* skeleton. Harvard University Press, Cambridge, pp 195–220

Walker A, Leakey R (eds) (1993) The Nariokotome *Homo erectus* skeleton. Harvard University Press, Cambridge

Walker A, Shipman P (1996) The wisdom of bones: in search of human origins. Weidenfeld and Nicolson, London

Case Study 16. Democratizing *Homo naledi*: A New Model for Fossil Hominin Studies

Abstract Analysis of hominin fossils generally requires access to the original material, but that lies scattered among museums around the world. New finds may sit for years inaccessible to scholars before they are formally published. Lee Berger has challenged this convention with his discoveries of two new species, *Australopithecus sediba* and *Homo naledi,* which he has made available to the field with rapid publication involving teams of both senior and junior scientists. These two species near the transition to genus *Homo*, join a series of recently recovered hominin skeletons. Each of them might tell us about the origin of humans, but each one seems to tell a different version of the story.

It has long been an ironic joke that there are more paleoanthropologists than fossil hominins. Disregarding isolated teeth, unaffiliated postcranial bones, and fragments, that remains close to the truth. Casts may or may not be available for purchase, but they are commonly expensive. Studying the real fossils requires travel around the globe, so that time and money become serious barriers for most anthropologists. It is unfortunate that they may face additional institutional or political obstacles to access.

The Closed World of New Hominin Fossils

When a new hominin fossil is discovered, by common convention it is the privilege of the finder to publish a description. Until that time, even if there has been a press release, other scholars are usually not permitted formal access to study or publish on it. Commonly, the finder releases an initial study. After that, if the fossil is important, the discoverer may decide who should perform a more comprehensive study. He (rarely she) may take that task on himself or delegate it to a colleague or student. A detailed description, analysis, comparative study, and evolutionary assessment usually require a number of years, and some descriptions

J.H. Langdon, *The Science of Human Evolution*,
DOI 10.1007/978-3-319-41585-7_16

have never been published. Only then is the fossil fair game for legitimate scholars to examine and write about. Practice varies, though most institutions are reasonably generous at this stage.

There is much to be said for this system. Each new fossil may add to or rewrite the sparse existing record. Ideally, large parts of our collections should be reassessed from the start with each major addition; but that is rarely practical. A thorough study establishes a record for all to consult and critique, although tracking down often obscure monographs, cost, and language may still constitute hurdles.

When *Ardipithecus ramidus* was first reported and named in 1994, it was a crucial find that extended the hominin fossil record 800,000 years further into the past, to about 4.4 million years ago. *Ardipithecus* was purported to be close to the ancestral line of the australopithecines and, ultimately, ourselves. As the initial material was announced in *Nature*, the team of anthropologists responsible for it, led by Tim White, made an even more impressive discovery of a partial skeleton. The bones were so fragile, they had to be protected in a plaster shell before they could be extracted from the sediments and taken to the laboratory in that condition. Such old fossils potentially could shed light on the origin of the hominin lineage and the initial evolution of such distinctive human traits as bipedalism. Anthropologists waited impatiently while the recovery and stabilization of the fossils proceeded slowly. The crushed skull and pelvis proved so fragile that virtual images of the fragments were created and manipulated on a computer to reassemble them.

As White's team conducted a thorough comparative and functional analysis of *Ardipithecus*, new discoveries were made. *Orrorin* (in 2001) and *Sahelanthropus* (in 2002), both at least 6.0 Ma displaced *Ardipithecus* as the oldest hominin, and the partial femora of *Orrorin* showed evidence of bipedalism. The still undescribed skeleton threatened to be an anticlimax.

White's team finally published their preliminary analysis in 2009, 15 years after its discovery. Their reconstruction was a great surprise and suggested that either the last common ancestor of humans and apes was not at all the chimp-like climber that was expected or that *Ardipithecus* was more distant from human ancestry than White claimed. As of this writing (2016), a more detailed study is ongoing and outside researchers have not been able to provide independent assessments of the original material. Although the *Ardipithecus* fossils present unusual problems, the access restrictions are not uncommon. For example, the partial skeletons from Dmanisi, described initially in 2007, have not been made accessible to outside researchers.

A New Business Model

The South African paleontologist Lee Berger has made a concerted effort to change this practice. His first chance came in 2008 when with his son he discovered australopithecine remains in Malapa Cave in the Cradle of Humanity region

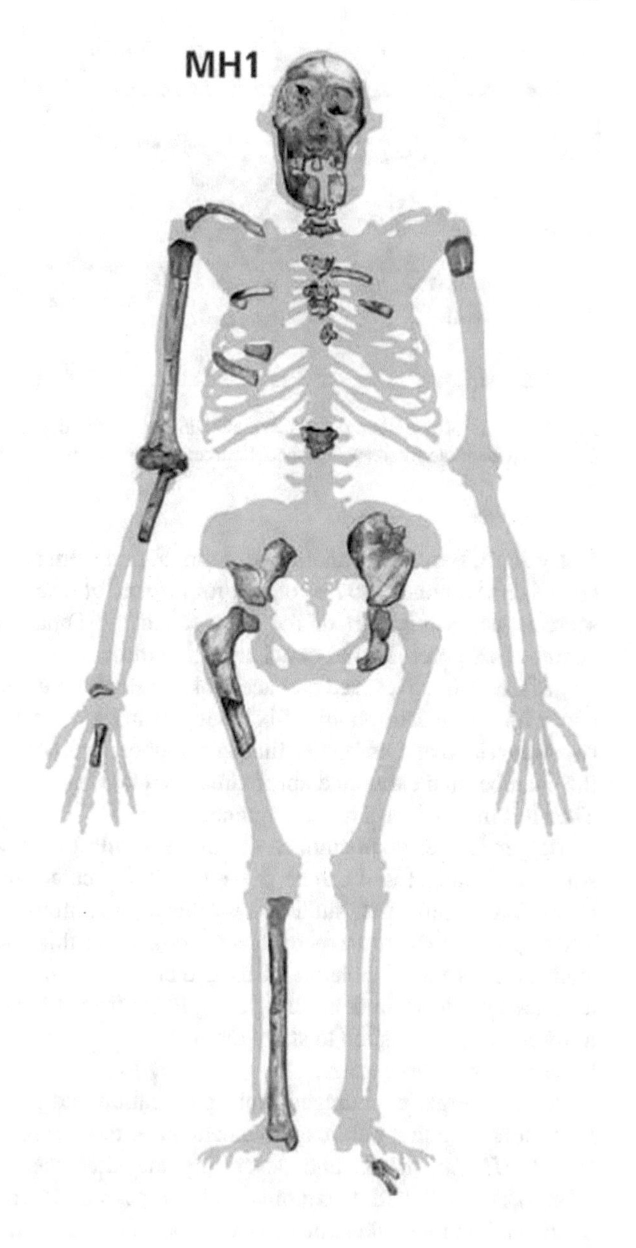

Fig. 1 *Australopithecus sediba* adult skeleton. Source: Peter Schmid, courtesy of Lee R. Berger. Creative Commons, with permission

near Johannesburg. The fossils represented two partial skeletons of a new species, *A. sediba* (Fig. 1). Rather than undertaking prolonged comparative studies, however, he pushed for a rapid publication so that other researchers could have access to the fossils. By dividing the work among a team of scholars with complementary expertise the work was accelerated. The initial description and naming appeared in 2010. A collection of five papers appeared in Science in 2011 and another seven in 2013.

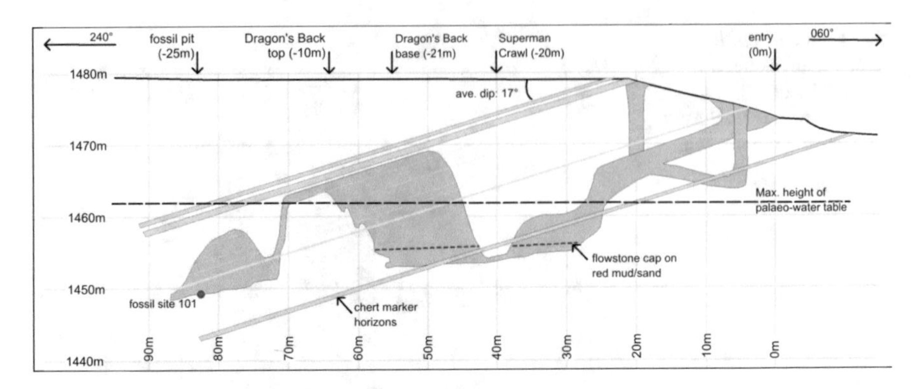

Fig. 2 Profile of the Rising Star Cave showing the restrictive passages leading to the Dinaledi Chamber where the fossils were found. Source: Creative Commons, with permission

By 2013, Berger had another sensational find from nearby Rising Star Cave. Two spelunkers responded to his request for information and reported finding bones in a scarcely accessible part of the cave called the Dinaledi Chamber. To reach the chamber, explorers had to crawl and slide through two extremely narrow passages (Fig. 2). Berger advertised on Facebook for small volunteers with scientific training and without claustrophobia. Six young women were selected. Operating with a two-way video and computer link to a support team outside the cave, they entered the chamber and excavated approximately 1550 bones from at least 15 individuals. They left most of the surface untouched.

Berger further democratized the process with this new discovery. The primary study of a major fossil can be the epitome of a career in paleoanthropology or its cornerstone; thus such studies are often monopolized by established academics whose grants fund the expeditions. Recognizing this, Berger worked through his team of 20 senior scholars and used them to recruit an additional 25 promising advanced graduate students and young PhDs from 12 countries. Then he convened a 6-week long workshop to study the material cooperatively and provided opportunities to jump-start careers.

Again, Berger encouraged rapid publication and provided support for his collaborators. The first public announcement of the fossils, two papers naming a new species, *Homo naledi*, and describing the site, was made in September 2015 (Fig. 3). A month later two more articles appeared, describing the remarkably complete hand and foot skeletons. As a further innovation, the bones of both *A. sediba* and *H. naledi* are being digitally scanned in three dimensions and the files are available for free download on the internet through Morphosource (at http://morphosource.org) so that anyone with a 3D printer can obtain his or her own copies at high resolution.

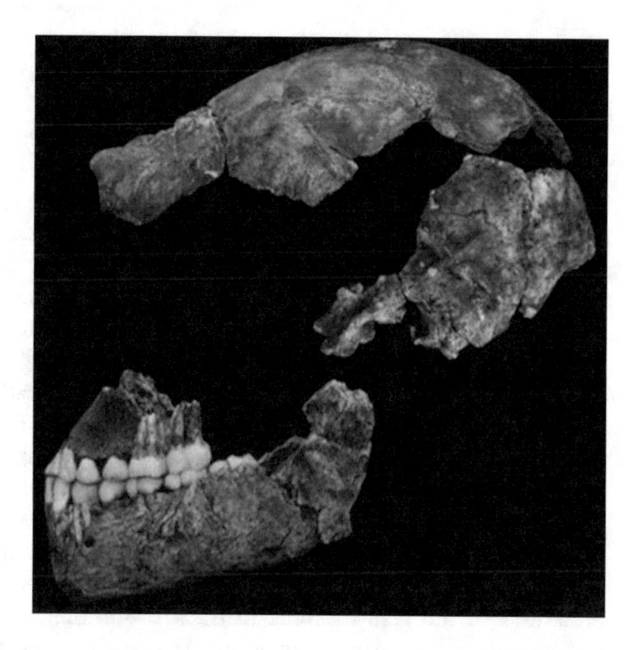

Fig. 3 *Homo naledi* skull. Source: Creative Commons, with permission

Berger was criticized in print by White for rushing to publication without a comprehensive study, and several anthropologists have challenged his interpretations. Berger's position is that additional comparative studies and/or alternative interpretations will follow anyway; but his strategy places the material in the hands of all researchers much sooner than the traditional approach.

Homo naledi and Mosaic Evolution

Both the site and the anatomy of *Homo naledi* are challenging our current understanding of human origins. The species position within our genus is yet to be determined. With a cranial capacity of 500 cm^3, it compares to the smaller specimens of *H. habilis* and the Dmanisi hominins. It therefore is likely to represent an extremely early branch.

Although the passages to the Dinaledi chamber may have been less constricted when *H. naledi* ventured there, no evidence of an alternative entrance has been found. The chamber presumably existed in absolute darkness since its formation, raising the probability that early hominins lighted their way with torches. No other species are represented in this tremendous collection of bones, aside from a few rodents and a single bird. This fact not only underscores the difficulty of past access, but also the inescapable conclusion that the remains had been deliberately placed there.

The inferred behavior, controlled fire and intentional deposition of the dead, is what we would associate with relatively advanced humans rather than a small-brained species.

Unfortunately, as of this writing, no date is available. The absence of other species rules out an easy faunal comparison, and Berger's team is waiting for consistent results from multiple dating techniques before releasing any information.

Anatomically, the skeleton poses additional puzzles. It is one more species, alongside *Ardipithecus*, *Australopithecus sediba*, *Homo* from Dmanisi, and *H. floresiensis* for which impressively complete postcranial material has become recently available. "A unique combination of primitive and advanced traits" is now a cliché to describe these species, but it is an accurate cliché nonetheless. Aside from *Ardipithecus*, all of these appear to inform us about the early history of genus *Homo* but they tell different stories. (Although much later in time, *H. floresiensis* is included in this comparison because of its generally primitive form and likely divergence from a very early point in the *Homo* lineage.)

In the evaluation of the position of a new species or fossil, brain size has always appeared to be an important consideration. Louis Leakey used a cranial capacity of 600 cm^3 as the defining boundary for genus *Homo*; and modern interpretations of the genus expect to track evolutionary change from that figure. Unfortunately for these expectations, the new species do not fit into an evolutionary progression or sort according to Leakey's standard. Instead, they indicate that diverse human species had occurred before brain size had increased to any significant degree (Table 1).

A second indicator of hominization is the relative lengthening of the lower limb. This is difficult to assess in absolute terms or compare among species because trunk length is rarely available and different bones of the lower limb are preserved in different fossils. However, when at least some long bones of both limbs are present, it is apparent that elongation occurs only in *Homo,* but in all of those species except for *H. habilis* (Table 2).

Table 1 Cranial capacities of fossil species showing morphologies expected of early *Homo. A. afarensis* and *A. africanus* are included for comparison

	Age	Cranial capacity (cm^3)	Comment
A. afarensis	3.6–3.0 Ma	387–540	
A. africanus	3.0–2.0 Ma	400–530	
A. sediba	1.977 Ma	420	$n=1$ adult
H. habilis	1.95–1.7 Ma	510–687	There is much controversy over which specimens to include in this species
Homo at Dmanisi	1.78 Ma	546–775	Species identity is uncertain
H. naledi	Unknown	513	$n=1$
H. floresiensis	100–60 Ka	417	$n=1$

Table 2 Selected features of the upper limb and hand of fossil species showing morphologies expected of early *Homo*. *A. afarensis* and *A. africanus* are included for comparison

	Shoulder and humerus	Curved phalanges
A. afarensis	Glenoid angled superiorly	Phalanges somewhat curved
	Lack of humeral torsion	Thumb relatively short
A. africanus	Lack of humeral torsion	Phalanges somewhat curved
		Thumb relatively short
A. sediba	Glenoid angled superiorly	Strong finger flexion
	Lack of humeral torsion	Phalanges somewhat curved
		Thumb longer than human proportion
H. habilis	Unknown	Phalanges somewhat curved
		Thumb of human proportions
Homo at Dmanisi	Extreme absence of humeral torsion	Unknown
H. naledi	Lack of humeral torsion	Extremely strong finger flexion
		Phalanges extremely curved
		Thumb of human proportions
		Unique keel on first metacarpal
H. floresiensis	Lack of humeral torsion	Unknown

The australopithecines, while bipedal, are understood as strong climbers, as evidenced by strong upper limb bones and longish somewhat curved phalanges. The arms were evolutionarily more conservative than the pelvis and foot. The shoulder (glenoid) joint was oriented somewhat superiorly and also ventrally, as indicated by the lack of torsion in the shaft of the humerus. This pattern continues in all of the early *Homo* species, while unique features may appear. For example, phalanges of the fingers of *H. habilis* possessed a slight curvature and in *H. naledi* that curve is greater than any hominin. In both *A. sediba* and *H. naledi*, the robusticity of the phalanges showed evidence of powerful flexor tendons. The first metacarpal of *H. naledi* has a unique keel on its ventral shaft for the attachment of muscles (Fig. 4). In *H. habilis* and *H. naledi* the thumb had elongated to modern proportions.

In the lower limb, the australopithecine ilium is widely flared and suggests different mechanics of balance than in modern humans. The femoral head is small and the neck relatively long. These features persist in *H. naledi* and *H. floresiensis* (Table 3). Although the foot of *A. afarensis* shows advanced traits, including a fully adducted first toe and partial shortening of the other toes, South African australopithecine feet are different, exhibiting a few very primitive features. The great toe of *A. africanus* is opposable and the heel of *A. sediba* is not weight bearing. In contrast, the foot of *H. naledi* is nearly modern, as is the partially known foot of *H. habilis* (Fig. 5).

When paleontologists compare a single ancestral species with a known descendant, there is an expectation that a fossil intermediate in time will be similarly intermediate in all anatomical features. However, these different species

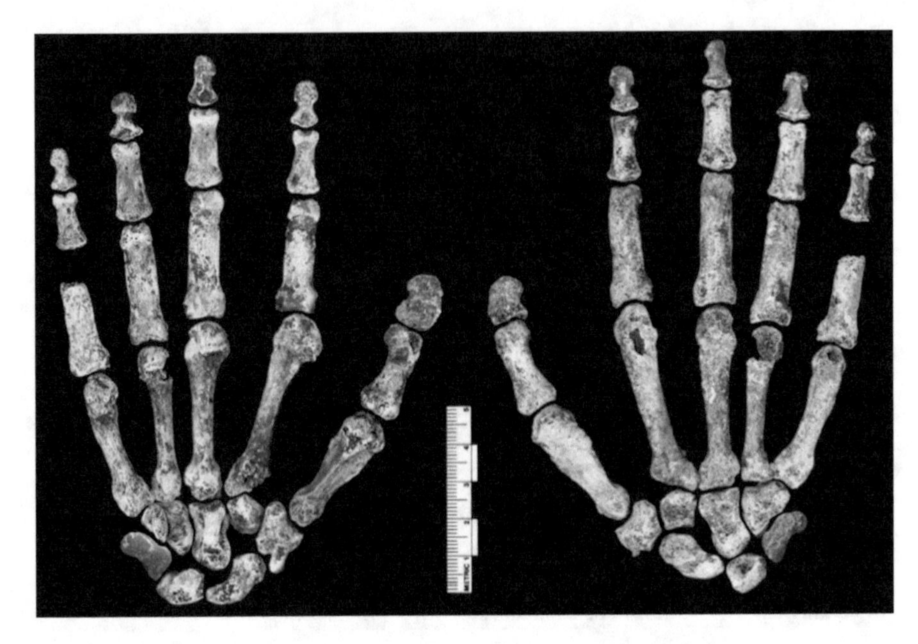

Fig. 4 The hand of *Homo naledi*. Source: Creative Commons, with permission

Table 3 Selected features of the lower limb and foot of fossil species showing morphologies expected of early *Homo*. *A. afarensis* and *A. africanus* are included for comparison

	Lower limb elongation	Pelvis and hip	Foot
A. afarensis	No	Flared ilium	Weight-bearing heel
		Small femoral head	Toes of intermediate length, phalanges curved
			Adducted first toe
A. africanus	No	Flared ilium	Divergent first toe
		Small femoral head	
A. sediba	No	Disputed iliac form	Non-weight-bearing heel
		Small femoral head	Adducted first toe
H. habilis	No	Unknown	Adducted first toe
Homo at Dmanisi	Yes	Large femoral head	Unknown
H. naledi	Yes	Flared ilium	Weight-bearing heel
		Small femoral head	Toes short, phalanges curved
			Adducted first toe
H. floresiensis	Yes	Flared ilium	Extremely long foot
		Larger femoral head	Toes of intermediate length, phalanges curved
			Adducted first toe

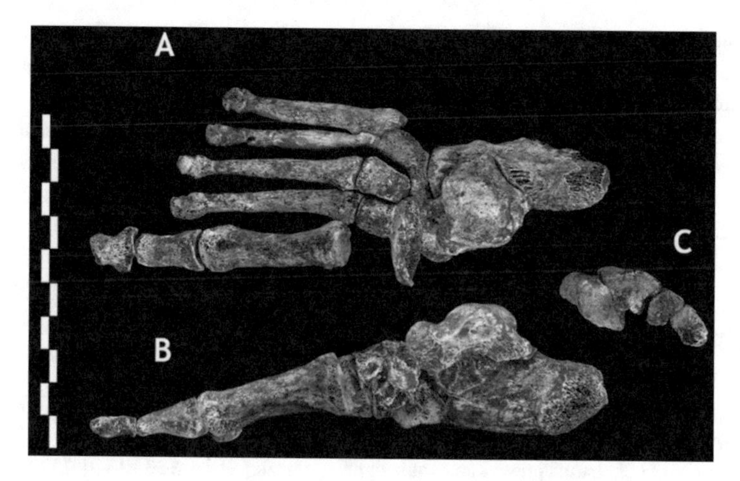

Fig. 5 The foot of *Homo naledi*. Source: Creative Commons, with permission

of hominin do not tell a simple linear story. Each body part has its own history and has evolved at a different pace and sometimes a different direction in each species to produce unique combinations of anatomy. This phenomenon is known as mosaic evolution. Perhaps the most important lesson it has to tell us is that human evolution is not linear, but the hominin lineage has produced a confusing array of side branches. There is not a single main trunk except in retrospect, and we are challenged at any time period to identify our ancestors among the known fossils or to even to know whether we have sampled our ancestor's species.

Questions for Discussion

Q1: If anthropologists are limited to the study of published photographs, descriptions, and measurements, how does that affect our understanding of the fossil record? What important information might be unavailable?

Q2: Why would any scientist prefer to share his major discoveries with a number of unknown young scholars?

Q3: Is it better for one team of scientists to spend years making a painstaking initial study of an important fossil or to speed to print and risk making errors that others may correct?

Q4: Compare the species in Tables 1, 2, and 3. How can we know which is our ancestor or have any confidence that one of them is?

Q5: Why is the upper limb so much more conservative than the lower limb?

Q6: Can a fossil with unique features that are not intermediate between an ancestor and expected descendant still lie in that pathway or must it represent a side branch?

Q7: As of this writing, no date is available for *Homo naledi*. How would our understanding of human evolution differ if it turned our to be contemporary with earliest *Homo* (about 2 Ma) or much more recent (a few hundred thousand years old)?

Additional Reading

Berger LR et al (2015) *Homo naledi*, a new species of the genus *Homo* from the Dinaledi Chamber, South Africa. eLife 4, e09560

Gibbons A (2011) Skeletons present an exquisite paleo-puzzle. Science 333:1370–1372

Harcourt-Smith WEH et al (2015) The foot of *Homo naledi*. Nat Commun 6:8432

Kivell TL et al (2015) The hand of *Homo naledi*. Nat Commun 6:8431

Shreeve J (2015) Mystery man: a trove of fossils found deep in a South African cave adds a baffling new branch to the human family tree. Natl Geogr 228(4):30–57

Stringer C (2015) The many mysteries of *Homo naledi*. eLife 4, e10627

Case Study 17. A Curious Isolation: The Hobbits of Flores

Abstract Of all the discoveries of fossil hominins, none has been more puzzling and unexpected than the "Hobbit" on the island of Flores in Indonesia. Anthropologists had no context in the fossil record in which to place it and had to grasp for analogies with a wide range of animals to make sense out of it. LB1 is an adult female that stands about 1 m tall with a brain capacity of about 400 cm^3—not large for a chimpanzee and unheard of for a healthy human. As recently as 60,000 years old, this little hominin was alive when modern *Homo sapiens* were spreading out of Africa.

Flores Island became a place of archaeological interest in 1998 when Mike Morwood announced a reliable date for stone tools that had been recovered from there beginning in the 1960s. Fission track dates indicated they exceeded 800,000 years in age. The published date, consistent with paleomagnetism, made these visitors contemporary with *H. erectus* in nearby Java. This island had been occupied long before by hominins who presumably had walked in from Asia when the sea level was much lower. Flores, on the other hand, has never been connected to the mainland. It is part of a region that lies between the Asian and Australian tectonic plates, and a deep-sea trench separates it from islands of the Asian plate. The islands are volcanic in origin, testifying to the dynamic geology of the region. To get from Java to Flores requires multiple crossings of deep channels and treacherous currents, including one of at least 25 km. This barrier has prevented most land animals from crossing. To the east, islands such as Java and Borneo are populated by fauna derived from Asian ancestors. To the west, Australia and New Guinea contain unique marsupials, reptiles, and flightless birds long isolated from the northern continents. This contrast was first noted by the naturalist Alfred Russell Wallace, and the line of demarcation is known as Wallace's Line (Fig. 1). Only birds and bats have easily crossed from one region to another, while a few other species have traveled by infrequent random events. Initial discoveries of ancient tools in 1968 and 1994 suggested early hominins were among those crossers, but that had been dismissed as too improbable. However, the additional finds in 1998 could not be ignored.

Morwood coordinated a more systematic search for evidence of human occupation. His team discovered the LB1 partial skeleton at Liang Bua Cave in 2003 and

J.H. Langdon, *The Science of Human Evolution*,
DOI 10.1007/978-3-319-41585-7_17

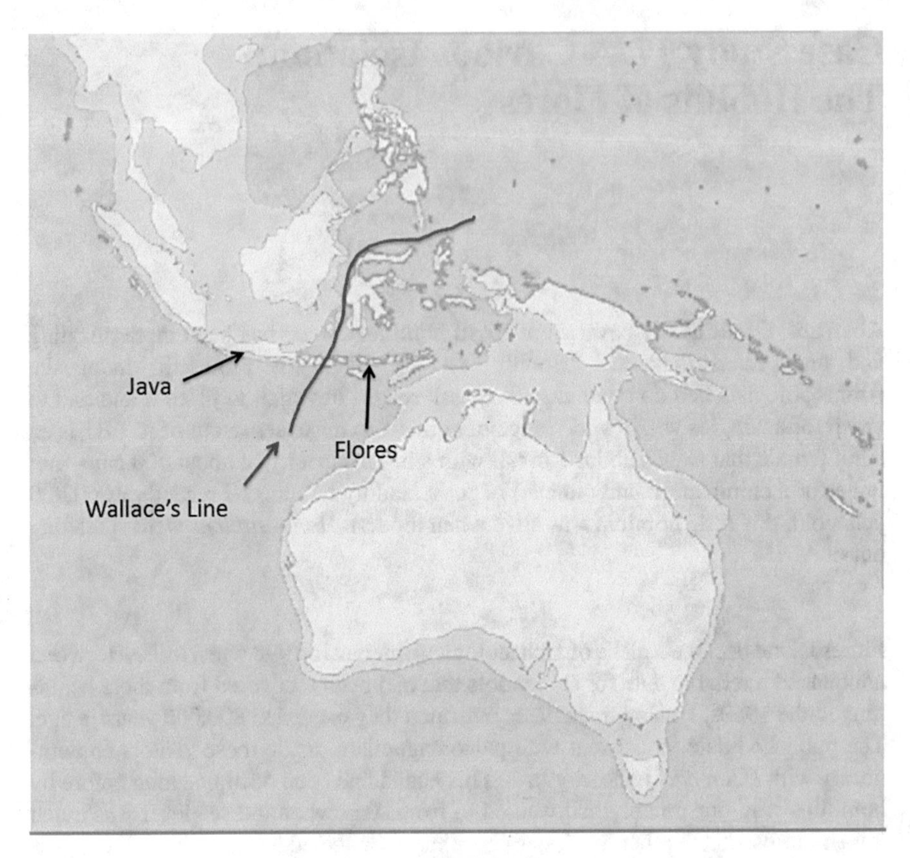

Fig. 1 The islands of Java and Flores lie on opposite sides of Wallace's Line, which separates the continental plates of Asia and Australia. Source: Creative Commons, Modified with permission

announced it to the world the following year, with the new species name *Homo floresiensis*. The editors of the journal *Nature* asked for a handier nickname and the field crew responded with the diminutive hero in a popular movie of that year, the "Hobbit." The skeleton was extremely fragile but included a complete skull, pelvis, and long bones of both upper and lower limbs, as well as parts of the hands, feet, and axial skeleton. The pelvis was interpreted as female. The femur was slightly shorter than the reconstructed length of Lucy, the smallest known australopithecine, and yielded a stature estimate of 106 cm (42 in).

LB1 was initially dated by both radiocarbon and thermoluminescence methods to about 18,000 years ago. An ESR date indicated the cave deposits extended as far back as 95,000 BP. In those deposits were more hominin teeth and isolated bones representing a minimum of 12 additional individuals. The most recent radiometric dating indicates that the bones and stone tools of *H. floresiensis* were laid down between 60,000 and 100,000 years ago.

The Shape of a Hobbit

The cranium has many primitive features, but lacks the jaw specializations and tooth size that would link it with australopithecines. It fits better with early *Homo*, and the logical comparison is with *H. erectus* from Asia. The heavy supraorbital tori (brow ridges) are familiar, but present as two arches over the eyes instead of the usual Asian pattern of a continuous ridge. The cranial vault rises minimally above the face but is more globular than in *H. erectus*. The face and jaws are moderately short and there is no chin, which only appears with modern humans. Published descriptions have numerous specific anatomical details of comparison, such as multiple mental foramina, that complete the mosaic of primitive, derived, and unique features. Unfortunately, it is difficult to distinguish which these are the result of its small size, and which are reliable shared derived characters that might tell us its affinities.

Smaller individuals of any lineage are likely to exhibit some predictable changes because they are small. For example, body mass diminishes more rapidly than body length measurements, so the bones and joints of a small individual will be more lightly built. Such size-related patterns of change are called allometry. It is easier to visualize allometric changes on the exaggerated "mouse to elephant" curve. The limb bones of a mouse are slender and fragile for their length compared to the robust bones of the elephant. Muscles are similarly disproportionate. Smaller animals also tend to have relatively larger brains and higher metabolism than expected from linear proportions. However, more precise changes that might accompany a reduction in size and stresses may be difficult to predict. Since no other archaic humans in the size range of *H. floresiensis* have been uncovered, scientists have nothing to compare its odd features to and cannot determine which are due to allometry.

The postcranial skeleton is no easier to decipher. Overall it appears very primitive. The pelvis has a marked lateral flare of the iliac blades, suggestive of *Australopithecus*. The femur has a longish neck, and the neck angle and head size are compatible with the primitive pelvis.

The feet are particularly odd. They are large, nearly 70 % the length of the femur. Modern people have a foot length about 55 % of the femur. The first toe was much shorter than the others. This resembles the primitive proportion of a chimpanzee toe, but the big toe is fully adducted (aligned with the other toes) in *H. floresiensis*, indicating that its condition is derived. The feet could not have grasped branches like an ape, but neither did it engage the first toe in a push-off that is compatible with rapid walking or running.

The humerus is somewhat more robust than human proportions would predict and does not show the torsion in the shaft that is typical of modern humans. This means that when the elbow is flexed, the hands would tend to spread more laterally. The clavicle is short and the shoulder probably was more protracted. These are traits shared with the earliest hominins, including australopithecines and Dmanisi *Homo*. Intramembral limb proportions (humerus to ulna length) are modern, but the upper limb is relatively long compared to the lower, showing proportions similar to Lucy.

The cranial capacity was estimated at 380 cm^3, compared with 1400 cm^3 for modern humans and about 900 cm^3 for *H. erectus*. In proportion to body size, it is comparable to chimpanzees or *Australopithecus*. The gross shape of the brain, reconstructed from three-dimensional CT scans, sorts it with *H. erectus*, and not with modern humans (including Pygmies), chimps, or *Australopithecus*. According to Dean Falk, the frontal region, which is small in *H. erectus*, is highly convoluted in LB1 in a unique way. Although the shape of the brain gives very limited information about its function, we must consider the possibility that LB1 brain reorganized in a different way from modern humans that makes direct comparisons of brain size less meaningful. The tool-making abilities and other evidence of behavior are more important indicators of intelligence.

Since the 2004 announcement of the discovery of *H. floresiensis*, an acrimonious debate ensued among anthropologists concerning its validity. Skeptics argue that LB1 is a pathological member of a local population of pygmy *H. sapiens*. There are scores of conditions that can interfere with development to produce a small brain, a condition known as microcephaly. Many of these have accompanying bone deformities. Trait-by-trait analyses can find humans matching most features that had been pointed out as evidence for a separate species status. Critics have argued for a much higher estimation of stature, closer to 120 cm (47 in.), and for a cranial capacity of 400 cm^3. However, they have failed to identify any one disorder that can account for all or most of the "abnormalities" in a single individual and would produced a brain as small as that of LB1. The discovery of additional specimens at Liang Bua and the extension of the date to before 60,000, preceding modern human presence in the region, makes the claim of a modern pathology even weaker. While those additional individuals are very incomplete—one is known by a single tooth—they present a consistent anatomical picture and confirm the small size. Another of the specimens is an infant whose limb bones are as small for modern babies as LB1 is for modern adults. This cannot be explained by postnatal stunting of growth. The controversies were intertwined with professional jealousies that impeded analysis of the bones and continued exploration at Ling Bua. Naming a new species assumes that a sustainable population once inhabited Flores, and continued searching is likely to turn up more remains.

Tools and Behavior

The tools found at Liang Bua are plentiful in number—over 3000 tools and flakes have been cataloged. The first descriptions reported sophisticated tool-making abilities, including the manufacture of blades, that compared favorably to the tradition brought in by later modern peoples. Such technology seemed conspicuously at odds with the small brain. However, any large collection of tools is likely to produce some that appear advanced. The great majority of the collection is less impressive and fits with the level of technology seen in the rest of Southeast Asia during the *H. erectus* period. The overall collection from multiple sites on Flores failed to show significant change over hundreds of thousands of years until the undisputed arrival of anatomically modern humans.

Liang Bua Cave has evidence of fire, from charred bones and fire-cracked rocks. It is not clear whether there was a constructed hearth or not, and what level of control was involved is unknown. *H. floresiensis* did hunt, butcher, and consume meat, as evidenced by animal bones accumulated in the cave with cut marks.

The large animals of Flores were not diverse. Because of its isolation, few species from Asia were able to colonize it. Most conspicuous in the cave were an extinct species of *Stegodon*—a relative of elephants—and giant rats, with deer and pigs also present. *Stegodon* had recolonized Flores after a pygmy species of it went extinct about 840,000 years ago. The largest predators on the island were varanid lizards, including the Komodo dragon that inhabits nearby islands today. There was also a giant 1.8 m (6 ft) species of stork, which was also carnivorous. *Stegodon* was described as another dwarfed species, mostly represented by juveniles, but has not been fully described. Even if dwarfed, it would have made a formidable prey, yet prey it must have been because *Stegodon* has not previously been found in a cave setting. The rats compared more to rabbits in size. Komodo dragons, on the other hand, can grow over 3 m in length, and an even larger species of lizard was also present on Flores. The humans were certainly both predator and prey in a strange ecosystem.

Island Dwarfing

Homo floresiensis leaves many unanswered questions. The bitter debate over pathology arises from the seeming improbability of any other explanation. If one assumes LB1 and the other skeletons were normal for their population, then island dwarfing is generally the best explanation for her small size. It is not uncommon for dwarf versions of normal species to evolve in isolation on smaller islands, such as Flores. *Stegodon* is a good example. The "normal" mainland species of *Stegodon* was a moderately small elephant about 2.5 m tall, as known from other parts of Indonesia. Some members of that species probably swam to Flores and lived there in genetic and evolutionary isolation, diminishing in size. Numerous other examples are known. Island populations of mammoth became dwarfed off the coast of California. Species of other elephants, hippopotamus, goats, buffalo, and even a dinosaur are known from island dwarf species. The primary argument against this model is that the brain size usually does not reduce proportionately with the body (but exceptions are known). The brain of *H. floresiensis* is therefore much smaller than one would expect.

To understand the dynamics at work behind these trends, we need a better understanding than we currently have of the controls of body size and life history strategy. Extended growth generally relates to delayed maturation. Larger body size has the advantages of more stored resources that permit an individual to withstand short-term environmental fluctuations. Larger individuals are better able to compete with members of their own species when size is an issue (e.g., direct confrontation). On the other hand, in the absence of competition, rapid maturation at a smaller adult body size permits an individual to reproduce more quickly. By shortening the generation time, it may be able to out-reproduce its competitors.

Three models have been put forward to explain island dwarfing. The first is simple genetic drift. If, by chance, the founding population contained smaller-than-average individuals than the parent population, the descendants of those founders would be expected to be smaller. Alternatively, if by chance larger (or smaller) individuals have more offspring, then the later population would consist of larger (or smaller) individuals. Genetic drift works purely by random overrepresentation of genotypes in the next generation and cannot explain why larger species consistently get smaller. Nor does it explain why small mammals, most commonly rodents such as the giant rats on Flores, but also birds and reptiles often get larger on islands. Genetic drift also cannot explain a sustained direction of change over time. Because *H. floresiensis* appears to be an extreme case of island dwarfing, we must assume it resulted from a sustained evolutionary trend over many generations, if not tens of thousands of years.

A second model assumes that a small island will have limited food and other resources for the population. Individuals who are smaller will need fewer resources and may have a survival advantage over larger individuals. The island can support a larger population of dwarfed animals than of the normal sized species. However, we know that selection favors successful individuals, not populations. If an individual is better able to control the resources it needs, it will thrive and reproduce at the expense of others. It is conceivable that larger individuals may have an advantage if such competition involves confrontation and intimidation. It is not clear that smaller body size would increase fitness in the individual. Faster maturation time to reproduction may be more important but that appears independent of resource limitations. Moreover, there is no prediction in this model to explain the increased size of small species.

The third model considers the effects of predation in determining body size. An animal may evade predators if it is very small and likely to avoid notice, or if it is large and/or fast enough to fend off a predator and escape it. When the species becomes isolated on an island, it is less likely that a large predator will be there also, since predators need much larger territories and prey populations to thrive. In the absence of predators, the selection to be small or to be large is relaxed, and other factors may predominate. Obviously *H. floresiensis* did face predators, but they were not the same species with which its ancestors had coevolved. Perhaps body size was less meaningful when facing lizards that today can prey on water buffalo. The trees or culture may have provided as much safety.

It is likely that island dwarfing and gigantism are best explained through some combination of these models. How well do they fit *H. floresiensis*? Humans are not usually thought of as prey, and it is surprising to think of human body size being determined by predation pressure, though the role of predators in hominin evolution is likely underestimated. The fact is, humans are subject to the same ecological rules and limitations as other animals. However, if *H. floresiensis* is a distinct species, such a perspective is the only logical way to understand it.

Questions About the Beginning and the End

Where did the early hominins of Flores come from? Postcranially, they are most similar to australopithecines or earliest *Homo*. Cranially, they are argued to align better with *H. erectus*, but there are similarities to other early *Homo* species as well. Could australopithecines or a very early species of *Homo* have wandered as far as Indonesia? There is no evidence that either of them ever left Africa. It is not clear where the Dmanisi hominins came from, either, though they are much closer to the East African hominins in time, space, and morphology. Could it be possible that *Homo* arose in Asia and then returned to Africa? *H. floresiensis* could therefore be a relic from this ancient lineage. It may not have dwarfed, but simply diverged before hominins became large. Unfortunately, there is no other evidence to sustain this version.

From present evidence it is also possible that some members of *H. erectus* rafted onto Flores by accident. Their descendants perhaps evolved rapidly because of the small size of the population. The very primitive features of the limbs may reflect an expression of retained primitive genes and/or the result of allometric changes during dwarfing.

This isolated population apparently persisted from 800,000 years ago until about 60,000 years ago. The hominin-bearing deposits at Liang Bua are capped at that date by the ash fall from a nearby volcano. Probably *H. floresiensis*, along with *Stegodon*, became extinct at that time. The eruption may have been the cause of those extinctions.

We don't yet know from where the "Hobbits" came or how they survived. We have yet to understand the process by which they became so small. We don't know what their tiny brains imply about their intelligence and behavior. There are so many questions yet to be answered, but we have known about the Hobbit for a short time. We can hope the mystery inspires further discoveries on Flores and neighboring islands that can tell us how long hominins lived there and how they evolved.

Questions for Discussion

Q1: The discovery of *H. floresiensis* was a surprise, to say the least. What preconceptions does it overturn?

Q2: The primitive upper limb traits of *Australopithecus* were interpreted as evidence of past or continuing importance of climbing. Is this likely to explain the anatomy of *H. floresiensis*? Does this mean all of its ancestors were also adapted to climbing in forest settings?

Q3: When we try to scale brain size with body size, is it valid to extrapolate from human populations?

Q4: *H. floresiensis* fits the pattern of island dwarfing. Should we be surprised humans are subject to the same ecological rules as other species? What ecological principles govern modern humans?

Q5: When we compare living species, intelligence approximately follows brain size. Should we estimate the *H. floresiensis* cognitive abilities by its brain size or is it likely to be an exception?

Q6: The arguments in this chapter sometimes use negative evidence. For example, there were no other hominins known on Flores; therefore, *H. floresiensis* must have made the tools; since only *H. erectus* is known from Asia, *H. floresiensis* must have descended from *H. erectus*. How strong is negative evidence and when should it be used to reject alternative hypotheses?

Additional References

Aiello LC (2010) Five years of *Homo floresiensis*. Am J Phys Anthropol 142:167–179
Brown P et al (2004) A new small-bodied hominin from the Late Pleistocene of Flores, Indonesia. Nature 431:1055–1061
Bunn A et al (2006) Early stone technology on Flores and its implications for *Homo floresiensis*. Nature 441:624–628
Jacob T et al (2006) Pygmoid Australomelanesian *Homo sapiens* skeletal remains from Liang Bua, Flores: Population affinities and pathological abnormalities. Proc Natl Acad Sci U S A 103(36):13421–13426
Larson SG et al (2007) *Homo floresiensis* and the evolution of the hominin shoulder. J Hum Evol 53:718–731
Moore MW, Brumm A (2007) Stone artifacts and hominins in island Southeast Asia: new insights from Flores, eastern Indonesia. J Hum Evol 52:85–102
Morwood M, van Oosterzee P (2007) The discovery of the Hobbit. Random House Australia, Milsons Point
Morwood MJ et al (1998) Fission-track ages of stone tools and fossils on the east Indonesian island of Flores. Nature 392:173–176
Morwood MJ et al (2004) Archaeology and age of a new hominin from Flores in eastern Indonesia. Nature 431:1087–1090
Morwood MJ et al (2005) Further evidence for small-bodied hominins from the Late Pleistocene of Flores, Indonesia. Nature 437:1012–1017
Wong K (2009) Rethinking the hobbits of Indonesia. Sci Am 301(5):66–73

Case Study 18. Neanderthals in the Mirror: Imagining our Relatives

Abstract Anyone who views hominin fossils has a desire to see them fleshed out. What did extinct species look like? How did they behave? Anthropologists and artists who try to answer these questions for us need quite a bit of license for their imagination, and often the results tell as much about modern humans as they do prehistoric ones. Of the extinct species, Neanderthals have been known and imagined the longest and have experienced the greatest number changes in their image. For the first half of the twentieth century, they were seen as primitive brutes next to civilized Cro-Magnon people. That image improved as perception of human nature took a turn for the worse. A new, humanized understanding of Neanderthals coincided with remarkable discoveries at Shanidar Cave.

One of the earliest published images of a Neanderthal is a terrifying sight (Fig. 1). A very ape-like visage looks menacingly out from the mouth of a cave. His face is pigmented like a gorilla with prognathic jaws and his body is covered with hair. In one hand he holds a club and the other a stone. An animal skull before him on the ground shows that these weapons in his muscular arms are lethal. This infamous depiction was printed in the *Illustrated London News* in 1909 following the discovery of the La Chapelle skeleton.

Boule's Neanderthal

Marcellin Boule's reconstruction of this fossil was the first to examine the full skeleton. In retrospect, there is much to criticize about his vision. Boule's Neanderthal was both stooped and stupid. The author appears to have gone out of his way to distance Neanderthals from modern humans and repeatedly compares them to apes. "The first vertebrae are more like those of a chimpanzee than those of a Man ... These peculiarities seem to indicate in the cervical region of the vertebral column either a complete absence of curves, or a slight curve, in the direction opposite to that in

J.H. Langdon, *The Science of Human Evolution*,
DOI 10.1007/978-3-319-41585-7_18

Fig. 1 Early depiction of a Neanderthal from the *London Illustrated News*, 1909

modern Man." Similarly, "It would seem as if the lumbar curve were less pronounced than in the majority of modern men." The sacrum bears "simian characters." The femoral shafts are compared to those of gorillas and chimpanzees. "Certain frictional surfaces [relating to the gluteal muscles] seem to indicate that the owners of these femora habitually maintained a bent posture." "[W]ithout being mechanically impossible, the total extension of the knee could not have been normal, and the habitual attitude must have been one of semi-flexion." The foot was also primitive, with a flat arch and an opposable first toe. "The [talar] head is much bent, denoting that the great toe was widely separated from its neighbors. The articular surface for the scaphoid points to a much depressed instep." "[T]he foot must have rested chiefly on its outside edge." The calcaneus is reconstructed without a lateral tubercle on the heel, making it look quite chimp-like. Neanderthals supposedly walked on the lateral side of the foot, somewhat pigeon-toed.

The La Chapelle cranium has a capacity of about 1600 cm^3, slightly greater than the modern average of 1450 cm^3. Boule dismisses this embarrassing statistic by making a functionally meaningless comparison to facial size and by considering this specimen to be the extreme end of variation in the Neanderthal population. "Thus there disappears, or is greatly lessened, the paradox seemingly indicated by the magnitude of the absolute volume of the La Chapelle skull, when due account is taken of the numerous signs of its structural inferiority." He concludes with reference to "the brutish appearance of this energetic and clumsy body, of the heavy-jawed skull, which itself still declares the predominance of functions of a purely vegetative or bestial kind over the functions of mind."

Boule's version of the Neanderthal as the antihuman fit closely with professional expectations and it infused popular culture for half a century. In addition to the primitive features that could be seen as more ape-like, the Mousterian culture of the Neanderthals consisted of cruder stone tools without the artwork, ornamentation, and inventiveness of later periods. Moreover, the site of Krapina in Croatia discovered in 1899 revealed evidence of bone breakage interpreted as cannibalism. The two leading scholars of human evolution in the first decades of the twentieth century, Boule in France and Sir Arthur Keith in England, rejected the known fossil hominins from our ancestry or even a close relationship. Brutish, violent, and dim-witted Neanderthals were depicted in museum displays, literature, and the cinema.

A few anthropologists refused to accept this portrait and argued that the bones indicated a fully upright carriage. This was systematically argued by William Straus and A. J. E. Cave who critiqued Boule's original reconstruction in 1957. Reexamining the La Chapelle skeleton, they identified pathologies, including arthritis, which caused the stooped posture, and they dismissed some of Boule's more imaginative attributions. Emphasizing the Neanderthal's modern stature and brain size, Strauss and Cave speculated that "if he could be reincarnated and placed on a New York subway—provided that he were bathed, shaved, and dressed in modern clothing—it is doubtful whether he would attract any more attention than some of its other denizens."

About the same time, other paleoanthropologists were reaching similar conclusions. Moreover, as the pessimistic Killer Ape image emphasized our violent nature, Neanderthals were advanced as peaceable contrasts. This idea was further developed in William Golding's 1955 novel *The Inheritors* and in other fiction. The extinction of the Neanderthals could now be interpreted as genocide at the hands of Cro-Magnon people arriving in Europe. A further sympathetic view derived from excavations at Shanidar Cave.

Shanidar Cave

The mouth of Shanidar Cave sits two-thirds of the way up a hillside overlooking a fertile valley in northern Iraq. The large entrance chamber and commanding view made it an attractive spot for early hunter-gatherers as well as later pastoral peoples. Ralph Solecki led a team of anthropologists there in 1950 in hopes of uncovering the past. He initially encountered a number of small huts and animal enclosures maintained by local herdsman. The floor was thick with accumulated ashes; centuries of debris; and the dung of cattle, sheep, and goats.

Solecki's team dug through 10 m of cultural debris. The top few layers produced Neolithic and earlier artifacts and 28 anatomically modern human burials. Further down were tools characteristic of the Levantine Aurignacian culture, associated with early modern people. Below that was a deep Mousterian layer indicating occupation by Neanderthals. The layers were dated by radiocarbon methods which show a fairly continuous deposition of soil, rocks fallen from the ceiling overhead, and cultural discards. The top of the Mousterian stratum neared the limits of radiocarbons dating—about 50,000 years. By extrapolating from the apparent rate of deposition, one can estimate the lowest cultural levels to have been 60,000–70,000 years old. Neanderthal skeletons were found at several levels within the Mousterian layer and apparently did not all live at the same time period, but possibly were scattered over tens of thousands of years.

Of the skeletal remains, seven were adults in different degrees of completeness and two were infants. The tenth, part of a young child's skeleton, was discovered much later among the animal bones in a museum collection. Although reasonably complete Neanderthal skeletons are fairly well known from Europe and Israel, including specimens of all ages, this constitutes one of the larger collections of Neanderthals from a single site. What makes them even more interesting is the history of injury and disease these skeletons reveal. Forensic anthropologists are trained to interpret recent bones for clues to pathologies and other events of life and death that affected them. Their skills can be applied equally well to ancient remains. The most thorough study, from which these summaries are taken, was made by Erik Trinkaus.

The Skeletons

The first standard questions to ask about a skeleton are the age and sex of the individual. Sex can often be determined from a mature skeleton by the shape of the pelvis, when present. Beyond that, males tend to have somewhat larger and more robust bones. These techniques are more difficult when the skeleton is incomplete, as many of these fossils are. Age determination is more challenging, even for modern humans. The degree of cranial suture closure and certain changes in the pubic symphysis are traditional forensic techniques employed by Trinkaus, but these give only imprecise answers. Other age-related changes include the extent of joint degeneration and tooth wear, both of which occurred faster than in modern people. A more recently developed technique for aging is based on the gradual replacement of bone that happens in the body. By counting rebuilt units of bone tissue, called osteons, in a histological section through the femur, it may be possible to obtain a more precise age.

Shanidar 1 is a relatively complete skeleton, though part of it, including the pelvis, was crushed in the ground and is beyond reconstruction. It is the most interesting of the skeletons, presenting a number of injuries, some of which may be related to one another. The individual is an adult male, based on the shape of the pelvis and the robust structure of the skeleton compared with Neanderthals from other sites. The age can be roughly estimated to a minimum of 25–30 years by the extent of fusion of cranial sutures. However, extensive wear of the teeth and other skeletal indicators suggest an even older individual, and Trinkaus concludes that an age of 35–40 is most reasonable.

The cranium is quite complete, but the reconstructed shape of the braincase is unusual. The frontal bone is flattened compared to that of other Neanderthals. In the absence of signs of relevant pathologies (there are plenty for other parts of the skeleton), Trinkaus suggests it was artificially shaped. Head binding to reshape the skull is a practice reported among many recent peoples around the world. If an infant's head is bound with a cord or strapped to a flat board, it will shape itself painlessly as it grows. The deformation is considered esthetic, but also serves to identify an individual with his or her culture.

The skull had suffered some injuries. Scars on the left frontal bone (forehead) indicate healed scalp wounds that had injured the periosteum and cut to the bone. More serious was a fracture of the frontal and zygomatic bones on the outside of the left orbit. The fracture is well healed, but the face is asymmetric as a result. This appears to have been caused by a severe blow to the side of the head and is likely to have injured the eye as well. Degeneration in the left jaw joint may have been related to this injury. Arthritic degeneration in the spinal column is more likely caused by age and wear.

The right upper limb is quite abnormal. All three bones present—humerus, clavicle, and scapula—are smaller than their normal counterparts on the left side. The humerus had been broken in its lower half at least twice. One of those fractures healed, but left the distal third of the shaft misaligned. A distinct callus, typical of a healing break, had formed over the surface. The lower fracture, just above the elbow, shows some reabsorption of bone around it, but no fusion with the lost tip of the humerus. It might have been an amputation, intentional or otherwise, since the rest of the limb is missing below that point. Overall the humerus is withered along its shaft. Strength and size of a bone is built and maintained by the actions of muscles attaching to it and to other forces acting on it. The state of this humerus, contrasting markedly with the normal left side, suggests long-term paralysis of those muscles and atrophy of the bone. The right clavicle, in addition to being shorter than the left, shows evidence of adjacent soft tissue injury. There is a callus of bone built up where an infection might have been harbored.

The right lower limb shows disease as well. There is severe degenerative joint disease affecting the knee and ankle and the bones and joints along the medial side of the foot. The fifth metatarsal on the outside of the foot was fractured and healed. The joint disease would have been painful and interfered with normal walking. Perhaps related to this, the right tibia is also deformed with a bowing of the shaft.

Trinkaus speculates on three possible scenarios of trauma to explain this suite of injuries. A crushing of the right upper limb, perhaps from a rock fall, might have caused the fractures and resulted in atrophy or interference with growth. The head blow may or may not have occurred at the same time. A second possibility is that a blow to the head caused brain damage that might have compromised the limbs on the right side of the body. The paralyzed upper limb would have then become more vulnerable to accidents and infections. Perhaps the useless limb had been amputated after further problems. A third possibility is that the injury around the clavicle damaged the brachial plexus, a network of nerves supplying the upper limb. Problems elsewhere in the skeleton would then have come from separate injuries. The body-wide degeneration of the joints may be related to the injuries combined with age and strenuous life. Eric Crubezy and Trinkaus have suggested another possibility, a disease called diffuse idiopathic skeletal hyperostosis (DISH). DISH is a condition of unknown cause that is relatively minor in its early phases but can become increasingly debilitating as it advances. Such speculations cannot be resolved from the evidence. Perhaps more remarkable than the presence of the injuries is the fact that the victim lived long after they occurred. In a time long before modern medicine, this crippled, one-armed, and perhaps one-eyed man, who was approaching old age by preindustrial standards, survived with his disabilities for many years.

Most of the skeleton and teeth of Shanidar 3 are present, but not the skull. He was a little over 40, as ascertained from extensive tooth wear, age-related change in the pubic symphysis, arthritic joint degeneration, and osteon replacement. Beyond normal aging, this individual suffered extensive degeneration in some of the joints of the right foot. Since the left foot appears to be normal, this was probably due to an injury. Another traumatic injury appears in the thorax. The left ninth rib was partially cut by a penetrating instrument, such as a stone blade. The rib above it is broken in the corresponding area and the distal fragment is missing, leaving uncertain whether or not it was also damaged. The ninth rib responded in attempt to heal itself, but the groove remained open, as though the object that caused the wound remained in place. From the extent of bone remodeling, the injury occurred perhaps a few weeks before death. Very likely it would have collapsed the lung, incapacitated the victim, and possibly caused his death. Attempts to reproduce such an injury experimentally tell us it was most likely caused by a thrown spear angling slightly downward when it struck its victim.

Shanidar 4 is a reasonably complete skeleton, but the bones are fragmented and fragile, making analysis difficult. Histomorphology of the femur suggests age in the mid-30s. This individual was probably a male. Joint disease was widespread in the body, but probably reflects normal age and use. A rib on the right side was broken and fully healed.

Shanidar 5 is represented by the skull and upper limbs and parts of the lower limbs. Probably a male around 40, this individual was relatively healthy, showing only slight arthritis and a healed scalp wound on the left side. Like Shanidar 1, who probably died at the same time, Shanidar 5 appears also to have had his head artificially shaped.

Shanidar 10 consists of the distal leg and foot bones of an infant between 1 and 2 years of age. An X-ray of the tibia shows a Harris line, a radio-opaque line that indicates a temporary interruption of growth. This might be caused by disease or malnutrition and is not uncommon in premodern populations. The position of the line indicates the insult occurred around 9 or 10 months of age by modern developmental standards.

The other skeletons are unremarkable for disease or injuries but are also incomplete. Shanidar 2 consisted of a flattened skull, much of the spine, and three bones of the limbs. The individual is a young adult and probably male. None of the bones show evidence of trauma, but there is minor joint degeneration in parts of the spine. Shanidar 6 was a young adult, possibly in her mid-20s, with relatively little tooth wear and only slight degeneration. The small size of the bones suggests female sex.

The skull and much of the limbs are present. Shanidar 8 is represented by a cranium, most of a foot, and a few other bones. They appear to belong to a small young adult. Two of the individuals, Shanidar 7 and 9 are young infants, probably less than a year old. They show no evidence of disease or trauma.

The assorted diseases, degeneration, and injuries attest to a physically demanding life that aged individuals quickly. Perhaps there is evidence here of interpersonal violence—the oldest such evidence, if it is true—but these wounds can be equally explained by hunting accidents and falls in the surrounding mountains. Life was hard and dangerous and often short. It was made bearable by the support of others.

The Social Context of the Bodies

These persons may have lived perhaps 20,000 years apart in time, as indicated by the depth of the skeletons in the floor of the cave. However, it is likely that several were buried at one time. Shanidar 1, 3, and 5 lay close together near the highest Mousterian layer and beneath or next to large rocks that had fallen from the roof of the cave. Solecki speculated that they had been killed in the rock fall. Shanidar 1 was lying on his back with arms crossed over the chest, suggesting a deliberate burial. On the other hand, Shanidar 5 was crushed and bent back so that the head was next to the pelvis. It is likely that he was left, or covered, where he lay. Shanidar 4, 6, 8, and 9 (an infant) came from a deeper layer of the floor. They were buried close together, again probably at one time, so that their bones became partially intermingled during excavation.

Were these individuals deliberately buried? Many excavators of Neanderthal skeletons in Europe during the 1800s and early 1900s claimed, or assumed, the individuals they uncovered had been buried, sometimes with elaborate ritual. The direct evidence was context, but that was destroyed during excavation. Field notes can be misleading, as some modern skeptical anthropologists prefer to doubt the interpretations of their predecessors. The hole in the ground where the skeleton lay may have been dug with reverence for the dead or simply was the easiest place to dispose of a decomposing corpse. When a body was placed in a fetal position, this may have prepared it for rebirth, or it may have made it easier to fit into the hole. The animal bones next to it may have been offerings to the dead or kitchen refuse. A circle of horns around the skull of a child may have been symbolic or may have been a random scatter misinterpreted by a credulous archaeologist. Such possibilities are difficult to resolve in retrospect. What is the appropriate skeptical position? Should we assume no spiritual life for the Neanderthals until we have undisputed proof, or should we assume they are like us until proven otherwise? These are questions that take us outside the realm of natural science. They speak of how we wish to view ourselves and our uniqueness on earth.

In this regard, Shanidar Cave produced one of the most tantalizing or perhaps misleading clues. As Solecki excavated the cave, he took routine soil samples and sent them back to the museum where they were forgotten. Eight years later Annette Leroi-Gourhan examined them for pollen as evidence of the paleoenvironment. She identified trees, shrubs, and grasses, as expected. When she viewed samples from around the Shanidar 4 skeleton, she found pollen from at least eight species of flowers, including hyacinth, bachelor's button, hollyhock, and groundsel. The pollen here occurred in dense clumps instead of the light random scatter that would be expected by the wind. She concluded that the body had been covered with a carpet of flowers. These flowers are present around the cave today and some are used as folk medicines. The Neanderthals may have buried their dead with the same esthetic—visual or olfactory—we use today. Or perhaps Leroi-Gourhan and Solecki allowed their imaginations to overinterpret a chance contamination by ancient, or modern, plants.

What can Shanidar reveal about the other members of the social group? For a crippled person or wounded invalid to have survived in the Paleolithic, other members of the group had certainly shared food and probably nursed his injuries. Is this genuine altruism—selfless sacrifices of effort or resources for others? It is commonly assumed that only humans with a moral consciousness are capable of true altruism. Perhaps the Shanidar cripple was still able to contribute wisdom or magic, so that he was not entirely a case of welfare.

The possibility of head binding to artificially reshape the skull is uniquely cultural feature. Body mutilation, including piercings, circumcision, scarification, tooth filing, and tattoos are observed in traditional and modern societies around the globe. Whether such modifications are to signal rites of passage or simply for beauty, anthropologists recognize a common underlying purpose. They permanently mark a person as a member of a culture or subculture and set them apart from outsiders.

The discoveries at Shanidar contrasted sharply to the brutish image from the turn of the century and contributed to a transformation in the way Neanderthals came to be viewed by many anthropologists: they became human. However, it is dangerous to carry cultural analogy too far. We legitimately approach evidence of burial with skepticism, demanding clear evidence; and the debate over Neanderthal nature continues. We cannot place ourselves inside Neanderthal society and understand the meaning of symbolic activities. Nonetheless it seems clear that the Neanderthals of Shanidar were not living as individuals occupying the same physical space, but as a social and economically interdependent community. That is a defining trait of humans.

Questions for Discussion

Q1: Why would Boule and other anthropologists have wanted to interpret Neanderthals as much more animal-like than ourselves?

Q2: We cannot tell whether the injuries at Shanidar were caused by violence or accident. What possible evidence might be able to answer that question?

Q3: How would we recognize a deliberate burial?

Q4: Does true altruism exist among modern humans? Or is our behavior always driven by considerations of self-interest, such as enhanced reputation or expectation of future reciprocity? Are these distinctions relevant when we look at fossil hominins?

Q5: Do our current reconstructions of Neanderthals and other extinct hominins reflect modern values?

Additional Reading

Boule M (1923) Fossil men: elements of human paleontology (transl. from 2nd French edition). Oliver and Boyd, Edinburgh

Cowgill LW, Trinkaus E, Zeder MA (2007) Shanidar 10: a middle paleolithic immature distal lower limb from Shanidar Cave, Iraqi Kurdistan. J Hum Evol 53:213–223

Crubézy E, Trinkaus E (1992) Shanidar 1: a case of hyperostotic disease (DISH) in the middle paleolithic. Am J Phys Anthropol 89:411–420

Dettwyler KA (1991) Can paleopathology provide evidence for "compassion"? Am J Phys Anthorpol 84:375–384

Solecki RS (1971) Shanidar: the first flower people. Knopf, New York

Straus WL, Cave AJE (1957) Pathology and posture of Neanderthal man. Q Rev Biol 32(4):348–363

Trinkaus E (1978) Hard times among the Neanderthals. Nat Hist 12:58–63

Trinkaus E (1983a) The Shanidar Neandertals. Academic, New York

Trinkaus E (1983b) Artificial cranial deformation in the Shanidar 1 and Shanidar 5 Neandertals. Curr Anthropol 23:198–200

Trinkaus E, Thompson DD (1987) Femoral diaphyseal histomorphometric age determinations for the Shanidar 3, 4, 5, and 6 Neandertals and Neandertal longevity. Am J Phys Anthropol 72:123–129

Case Study 19. Leaving Africa: Mitochondrial Eve

Abstract For more than a century paleoanthropologists have been arguing over the relationship between the fossils and regional populations of modern humans. While the fossil record should be able to identify the time and place of the first modern people, it remains too incomplete for such a task. However, all of this argument and speculation was declared to be at an end with the publication of a 1987 paper that claimed geneticists had found the answer: We are all descended from a woman who lived in Africa about 200,000 years ago. The study by Rebecca Cann, Mark Stoneking, and Allan Wilson was so creative and sure of itself, and so elegant in its reasoning that it appeared to be definitive. Unfortunately it was also subtly simplistic, leaving the next generation of researchers to improve on it, and create a much messier picture. Uncertainties do remain about the mitochondrial ancestor, and that study describes only one dimension of human genetic evolution; nonetheless, "Mitochondrial Eve" revolutionized the way anthropologists think about human populations.

By 1.8 Ma, populations of early *Homo* were leaving Africa to colonize the Old World. Their descendants, *Homo erectus* showed up in Java and China. Under the name *H. heidelbergensis* humans occupied Europe a million years ago, apparently evolving later into *H. neanderthalensis*. Still others stayed home in Africa. The trend in recent decades is to stress the differences among these populations and place them into different species. All of these groups have left a small number of fossils for us to ponder. From which group did modern humans evolve?

In the second half of the twentieth century, paleoanthropologists debated two competing models. According to one, all fossil populations belong to a single continuous lineage that led to *Homo sapiens*. Regional populations interbred with their neighbors, but not sufficiently to keep from acquiring genetic differences through mutation, drift, and selection. Neanderthals are the direct ancestors of today's Europeans, "Peking Man" (and Woman) gave rise to Asians, and so on. Inherent in this model is the assumption that racial differences deeply rooted in time define our species. When anthropologists rejected the concept of

J.H. Langdon, *The Science of Human Evolution*,
DOI 10.1007/978-3-319-41585-7_19

race for biological and political reasons in the 1960s, this model fell out of favor. However, that judgment disregarded intriguing similarities and possible transitional fossils that seem to link past and present regional populations. More recently, some anthropologists have defended this model under the name "Multiregional Hypothesis."

The competing view acknowledged continental divergence of populations in the Middle Pleistocene that led to different species. Modern humans, it argued, could only have evolved from one of those; the others were evolutionary dead ends. When one species eventually developed competitively superior traits, it multiplied and spread out across the hemisphere, replacing less adaptable archaic peoples with modern humans. The "Replacement Hypothesis" looked to Africa for the origin of modern humans because advanced skeletal traits and behaviors seemed to appear there first; hence, it became known as the Recent Out of Africa model. It could also answer more easily the reasons why modern anatomy does not show up in Europe until tens of thousands of years after it appears in Africa.

Given the imperfection of the fossil record and the ambiguity of tracing descent across hundreds of thousands of years of sporadic, incomplete, and variable fossils, it seemed unlikely that fossils alone could resolve this debate. The solution would have to come from a different and independent source of evidence.

The Special Properties of Mitochondrial DNA

Rebecca Cann was a student of Allan Wilson, one of the authors of the molecular clock. The clock, as formulated in 1967, depended on counting accumulated genetic changes in distinct species. It could not be applied to create a phylogeny of modern human populations because they had been interbreeding and exchanging genes throughout human history. Cann and her coauthors found a way around this problem by using mitochondrial DNA.

Between one and two billion years ago, an oxygen-using bacterium invaded a larger cell. Perhaps it was a parasite; perhaps it was a meal. Either way, the smaller organism stayed and made itself indispensible. Many bacteria find free oxygen lethal because of its ability to react with and degrade DNA. This symbiote not only provided some protection by metabolizing oxygen, but also created a more efficient recovery of energy that could be used by the host cell. The cells of all plants, animals, fungi, and complex one-celled organisms contain the descendants of this visitor, which we call mitochondria. A mitochondrion is the organelle in which aerobic respiration takes place to capture energy from the breakdown of other molecules. The evidence of its origin lies in the fact that the mitochondrion retains its own cell membrane and its own DNA. Although some of the original mitochondrial genes have moved to the nucleus of the cell, some remain in the organelle and replicate as the mitochondrion reproduces. In humans, the mitochondrion still contains 37 genes arranged on a circle of DNA 16,569 base pairs long.

The mitochondrion and its DNA can multiply within the larger cell independently. More importantly, they do so without sex or exchange of genes. When a sperm fertilizes an egg, the body of the sperm stops at the egg cell's membrane and injects its chromosomes. The resulting offspring carries chromosomes from both mother and father, but its mitochondria only come from the mother. If it were not for occasional mutations, the mitochondrial DNA (mtDNA) of every person would be an exact copy of that of his or her mother. In theory, any person could trace that same pattern of mtDNA through generations of mothers and grandmothers indefinitely. However, mutations do occur, so when the mtDNA of different people is compared, there is variation—just as nuclear genes vary from person to person.

Mitochondrial Eve

The mtDNA of any given population or species can be traced to a single ancestral individual. In the time since that individual lived, many genetic variations were created and many went extinct. She may have had contemporaries in the population, but their mitochondrial lineages are all dead ends. How do we know this?

An analogy can be made to tracking family names. Imagine passengers on a ship marooned and permanently isolated on an island. They marry and have children among themselves and so on through future generations. At the beginning, they have 50 different last names that are passed on from father to son without exception in the American tradition. If in any generation a father has no son, his name will be lost from the population. Given enough time, this will happen to all names until only one is left.

The human species is a finite population. Every generation, some women have no children and some have only sons. Like our marooned passengers, each woman faces the chance that her mtDNA lineage will be lost in any given generation. Over time, only one mtDNA lineage will be left. The length of time will depend partly on the size of the original population and partly on chance, selection, and similar factors. It may take several generations or millions of years, but given a finite population and infinite time, it will occur. Here is where Cann and colleagues were able to construct a molecular clock. If we know how fast mtDNA changes and we know how great differences are among people, we should be able to estimate the time of divergence since the last common ancestor. If we could survey enough human beings, we should be able to calculate when the last woman lived who was the ancestor of all of our mitochondria. That ancestor might not be a member of *Homo sapiens* and may have lived long before hominins existed, but we can be certain that such an ancestor did exist.

Cann collected mitochondrial DNA from placentas from 147 women giving birth around the world. Her sample included 46 Caucasian Americans, 34 Asians, 21 aboriginal Australians, 26 aboriginal women from New Guinea, and 20 women of

African descent (18 of whom were African Americans). She examined only a small segment of the mtDNA, but found 134 different genetic sequences. A computer program calculated the most parsimonious "family" tree that could explain observed diversity with the fewest mutations. This allowed her to estimate the number of mutations that had occurred since the last common ancestor.

To calibrate her clock, Cann turned to the archaeological record. By the information available to her at the time, the first people arrived in Australia about 40,000 years ago. The mutations that set them apart from other peoples were thus accumulated in the past 40,000 years. Similarly, she assumed New Guineans have been isolated for 30,000 years and Native Americans for 12,000 years. This indicated that the region of DNA she examined changed by 2–4 % per million years; therefore, the last common ancestor lived between 290,000 and 140,000 years ago. She referred to this universal mother as "Mitochondrial Eve."

On the tree created by the computer, each set of mutations divided the subjects into two groups, those individuals with and without the new mutations. When Cann examined the divisions created by what was recreated to be the earliest mutation, both had representatives from Africa. One group only contained Africans and the other contained some Africans and all of the non-Africans (Fig. 1). She interpreted this to mean that the first mutations occurred in an African population and that the common ancestor herself was African. As she followed later subdivisions of the population, the data allowed her to extrapolate the times when different populations diverged from the others. The exodus from Africa occurred anytime after 180,000–90,000 years ago. Asians diverged from other peoples between 105,000 and 53,000 years ago, Australians 85,000–43,000, Europeans 45,000–23,000, and New Guineans 55,000–28,000. If all modern people belonged on this tree, there was no room for descendants of Neanderthals or other archaic humans.

Cann's publication had profound implications that rocked the field. She and her colleagues concluded, "Thus we propose that *Homo erectus* in Asia was replaced without much mixing with the invading *Homo sapiens* from Africa." Once again paleoanthropologists were being told that genetics made their observations of fossils irrelevant. Rather than worrying whether Neanderthals and *H. erectus* were related to later humans, they should be asking how it was that *H. sapiens* so quickly and so completely outcompeted all other archaic species. Was it language, intelligence, superior technology, or better social organization? Opponents scornfully referred to the hypothetical migrants as "killer Africans."

Still other implications became apparent: Could all the human genetic variation we observe today have arisen only in the past 100,000 years? Skin color, body shape, and the rest must be therefore able to change very fast. Could this be the first time in known human history that an invading population did not have sex and babies with the resident women? If not, the species barrier between them must have been unimaginably secure.

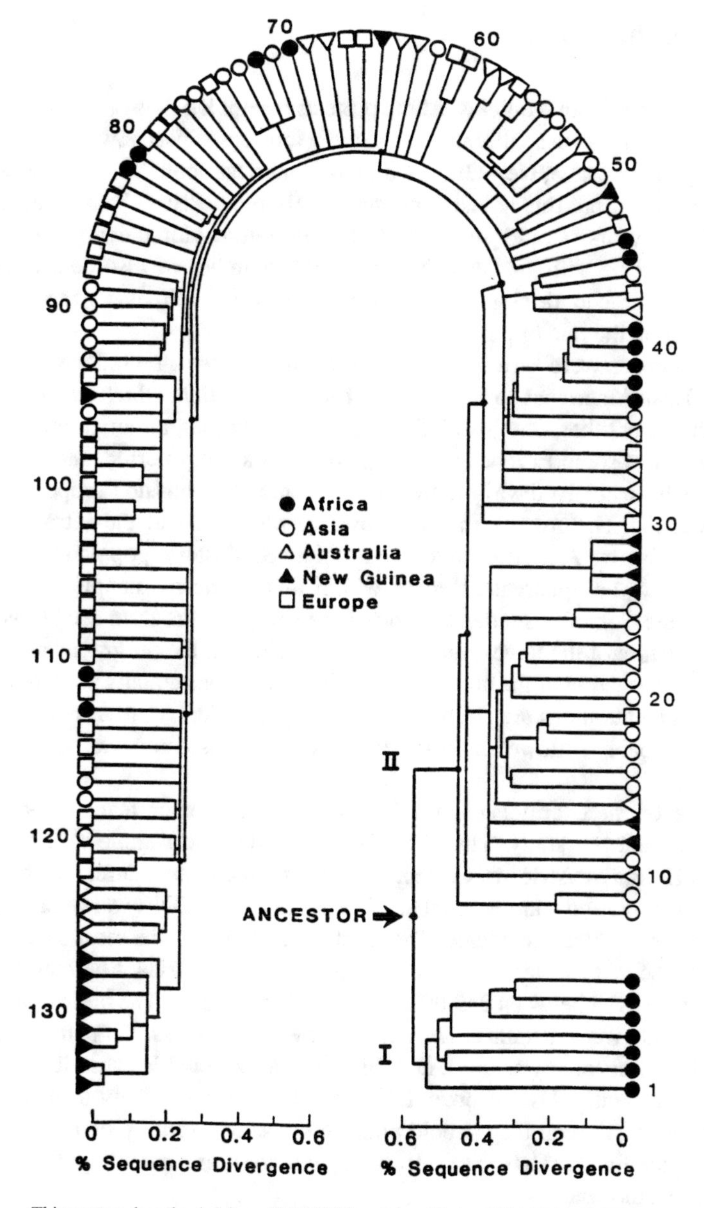

Fig. 1 Phylogenetic tree constructed from mtDNA sequences of 134 modern individuals. The deepest division, representing the greatest number of mutations, separates the first seven individuals from the rest. Because all of the first clade and some members of the second clade have African origins, it is most likely that the last common ancestor of this sequence lived in Africa. Reproduced from Cann R (1988) DNA and human origins. *Annu Rev Anthorpol* 17:127–143 with permission

Adjusting the Model

Scientific criticism came quickly. Many protests resembled those leveled at the first molecular clock. Was the calibration accurate? Anthropologists now believe humans first arrived in Australia, New Guinea, and the Americas about 25–50 % earlier than Cann's estimates, but this adjustment would only move "Eve" back a few tens of thousands of years. Could selection have interfered with the rate of change? Selection must be acting—we now recognize a number of disorders caused by mutations to mitochondrial genes—but we are not yet able to assess the full extent of its impact on the rate of population divergence.

Two criticisms were serious enough to require the work be redone. Cann used mostly African Americans to sample the African population, which she found more diverse than any other. She justified this by assuming that, although most African Americans have some European ancestry from the slavery period, such couplings would only have involved white males and black females; thus no European mtDNA was introduced. Historically that is known to be false, and the introduction of European or Native American mitochondria into the historic population may well have increased the apparent diversity within this "African" sample. The second problem came from a misunderstanding of the computer program used to generate the tree. Because of the large quantity of a data involved, the program was set up to sample only a fraction of the possible trees. Other solutions equally parsimonious or even more parsimonious are likely to have been overlooked. Furthermore, although evolution probably followed a parsimonious course, this can never be determined for certain.

Because of these criticisms, Linda Vigilant and another team working with Wilson repeated the study. This time they expanded the sample to 189 people, including 121 native Africans. Chimpanzees were used as an external calibration of the rate of change and a larger number of base pairs of the DNA were examined and all possible trees were considered. The results were similar. The most parsimonious solution to the last common ancestor still placed her in Africa. She now was estimated to have lived between 249,000 and 166,000 years ago. The basic conclusions of Cann's study were sustained. Once more, however, only a small part of the mitochondrial chromosome was used, and this one was selected because it was known to mutate frequently. Use of other DNA regions would be likely to give slightly different results, though they should point to the same overall picture. Because of this and uncertainties relating to selection, parsimony, and sampling, these studies cannot tell us the whole story.

Who Was Mitochondrial Eve?

Cann's choice of the term "Mitochondrial Eve" was unfortunate because it incorrectly associates this hypothetical woman with the first couple of the human species. Since a species is defined as a population evolving through time, that

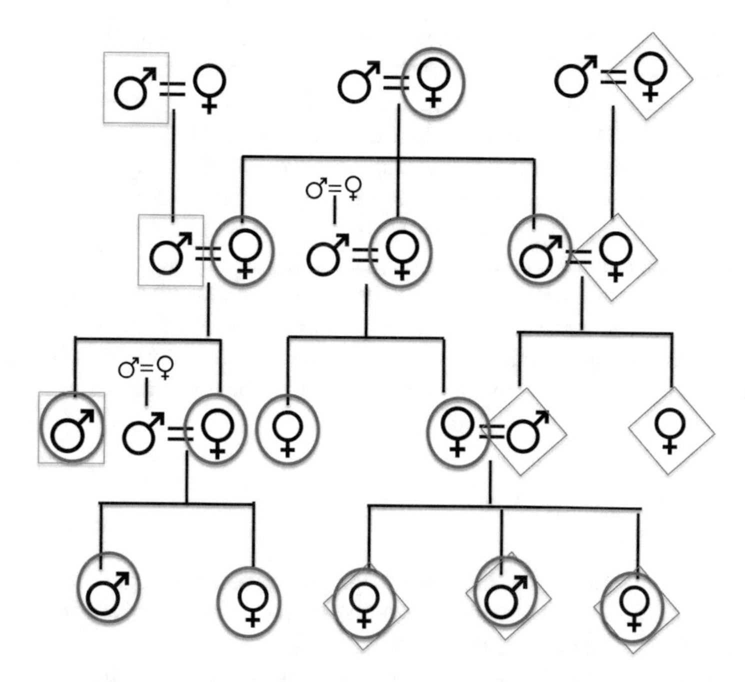

Fig. 2 This pedigree shows the different inheritance patterns for mtDNA (*circles*) transmitted only from mothers to offspring, Y chromosome (*squares*) transmitted from fathers to sons only, and nuclear DNA (*diamonds*), which have a probability of transmission to any offspring. A given mitochondrial genome or Y chromosome may be lost when a given generation fails to have daughters or sons, or may displace competing genes in the population as shown. However, that does not prevent nuclear genes from being transmitted such that later generations can still claim descent from a large number of ancestors in earlier generations

concept is not even meaningful. The calculation of the common ancestor tells us nothing about taxonomic groupings, and she could have lived long before or long after the start of our species. In fact, however, 200,000 years ago is reasonably close to when near modern humans first appear in the fossil record.

"Eve" was not the only woman of her time to contribute genes to modern populations. Our nuclear genes might have come from anyone in her breeding population, including women who had only sons, or from any other population that interbred with descendants, without affecting the mitochondria (Fig. 2). That includes Neanderthals and *Homo erectus*. Nuclear genes are rearranged through sexual reproduction. While they may be weeded out as well through selection and genetic drift, those are independent processes.

Yet another perspective was offered by John Hawkes: Cann's data are consistent with a genetic "sweep." A sweep describes strong selective pressure that favored only a single variant to survive among all the possible mitochondria in a breeding population. Rather than viewing this as a competition among regional populations, we may think of it as competition among genes in a single global breeding population. Perhaps Eve's mitochondria produced energy more efficiently to that her daughters had an advantage over other people. At the same time, any of her contemporaries among archaic peoples in Europe and Asia may well have contributed nuclear genes. If this is true, Cann's data do not resolve the original debate at all.

The story of Mitochondrial Eve is a history only of the genes in a part of the mitochondrion. She was a real individual whenever she lived; however, one could perform a comparable study on any gene or segment of DNA and it would tell a different story. It might take us much further into the past or to some other part of the globe. Many molecular anthropologists today are pursuing those pieces of evidence. There is an expectation that as the histories of enough genes are traced we will come closer to understanding the origin and movements of modern people. Because of sexual reproduction, however, there happen to be few other pieces of our genome whose story can be told as simply as that of the mitochondrion.

Questions for Discussion

Q1: Must the date at which one population diverges from the rest (e.g., Native Americans) correspond with the date when that ancestor arrived in the region with which her descendants were associated?

Q2: What has happened historically when two human populations that are morphologically, technologically, and culturally distinct come into contact? Can such historical examples help us to understand prehistoric events?

Q3: What genes are inherited only from the father? Could we calculate an Adam? If so, why might he be found to have lived in a different time and place? How is it that male and female genes have different histories?

Q4: What other independent sources of information besides bones and genes might help us trace the origin of modern humans?

Q5: Did the introduction of genetic studies qualitatively change the development of paleoanthropology, or would continuing fossil discoveries eventually have led to the same conclusions?

Additional Reading

Cann RL et al (1987) Mitochondrial DNA and human evolution. Nature 325:31–36

Hawks J (2008) Selection on mitochondrial DNA and the Neanderthal problem. In: Harvati K, Harrison T (eds) Neanderthals revisited: new approaches and perspectives. Springer, Dordrecht, pp 221–238

Relethford JH (2003) Reflections of our past: how human history is revealed in our genes. Westview, Boulder

Vigilant L et al (1991) African populations and the evolution of human mitochondrial DNA. Science 253:1503–1507

Case Study 20. The Neanderthal Problem: Neighbors and Relatives on Mt. Carmel

Abstract Mount Carmel in northern Israel lies on the route of any people traveling between Africa and Eurasia along the Mediterranean. It has been a sacred site and a refuge for fugitives from the world or merely from the law; but among its pilgrims today are paleoanthropologists. Numerous caves and archaeological sites on and near the mountain bear witness to the paleolithic cultures and hominin populations that have inhabited the region over hundreds of thousands of years. They include Neanderthals and other archaic peoples who probably missed one another by not so many thousands of years. The sites help us to perceive the differences among these types of humans as more ecological than technological.

The Neanderthal Problem

The first recognized Neanderthal specimen was discovered in Europe in 1856. After Neanderthals became understood as a distinct population, anthropologists faced the task of unraveling the relationship between them and anatomically humans. For the first part of the twentieth century, the tendency was to exclude archaic-looking fossils from direct human ancestry, especially those with smaller brains, and relegate them to side branches of the tree. This is how *Homo erectus*, *Australopithecus*, and Neanderthals were originally received. Continued discovery of fossils eventually made it clear that Neanderthals were the sole inhabitants of Europe until about 45,000 years ago when anatomically humans arrived on the scene. Both groups had similar brain sizes, but modern humans introduced a more complex and dynamic material culture. The questions of how these two peoples were related, how they interacted, and what happened to the Neanderthals became known as "the Neanderthal problem." An imaginative variety of solutions have been proposed:

- Neanderthals evolved directly into anatomically humans.
- Neanderthals went extinct because of climate change and anatomically humans moved into a depopulated continent.
- Modern humans outcompeted Neanderthals through superior technology, intelligence, and language.

J.H. Langdon, *The Science of Human Evolution*,
DOI 10.1007/978-3-319-41585-7_20

- Modern humans clashed violently with Neanderthals and carried out a prehistoric genocide.
- Neanderthals and anatomically modern people interbred, but the greater numbers of invaders genetically swamped the smaller Neanderthal population.
- Neanderthals survived (and perhaps still do) as a relic population in remote areas of the world.

Since the fossils have not changed, these different answers largely reflect changing attitudes about human nature.

Similar debates took place among archaeologists. The Neanderthal tool culture, known as the Mousterian, persisted for a long period with very little technological advancement. As anatomically humans arrived, they brought with them much innovation and a culture that was changing rapidly. Modern humans in Europe are associated with not one but many different tool traditions. The contrast between the two patterns is so great that the boundary is considered the transition from the Middle Paleolithic (Mode 3) to the Upper Paleolithic (Mode 4) grades of technology.

Current genetic research, additional evidence, and refined dating have improved understanding of the problem, but still leave unanswered questions. Modern humans entered Europe by 45,000 years ago and rapidly expanded from east to west in a period of about 5000 years. They appear to have originated in Africa. The Neanderthal population contracted as rapidly, with the last holdouts surviving in the far corners of Europe, Spain, and Portugal in the West and the Caucasus Mountains in the East. Modern Europeans entered the stage with a tool culture known as the Early Aurignacian, which was distinct from the Mousterian but not much more sophisticated. However, both cultures began to evolve quickly. The Late Aurignacian invented new implements that made the tool kit more varied, more specialized, and more efficient. Upper Paleolithic peoples acquired better weaponry, art, and complex ritual. The Mousterian also changed as the cultures overlapped, with late Neanderthals in the west apparently adopting a number of Upper Paleolithic innovations, including blades and ornaments. Several local-derived cultures survived briefly, including the better known Châtelperronian, but died out with the Neanderthals themselves.

The picture in the Near East is more complex. Israel and neighboring areas lie on the obvious pathway by which Africans would reach Europe, but there was no single, dramatic migration. Instead, modern or near modern humans appear there much earlier, yet the tool cultures remain comparatively static. Key to understanding these patterns are the caves on and around Mt. Carmel.

The Caves

Mount Carmel in northern Israel has a history both sacred and secular stretching back to the origin of *Homo sapiens*. The "mountain" is actually a 39-km long mountain range that contains sacred sites for the Egyptians, Canaanites, Hebrews, and Romans, but has housed hominins from the depths of prehistory (Fig. 1). Four caves

Fig. 1 Mount Carmel, Israel. Source: Public domain

on Mt. Carmel and others in the region have preserved the remains of Paleolithic humans. Three contain Neanderthals and two near modern *H. sapiens*. These and nearby sites present a very complete record of technological sequences from the Lower Paleolithic to the present.

The first important and systematic excavations into the prehistory at Mt. Carmel were conducted by one of the first women in the field, Dorothy Garrod, between 1929 and 1934. She excavated three caves near the western end of the mountain, close to the Mediterranean coast. Two of them, el-Wad and Mugharet et-Tabun contained a long and nearly continuous deposition of soil and tools from the Lower Paleolithic Acheulean to the Upper Paleolithic. At Tabun, she discovered remains of at least two individuals—an isolated mandible and the skeleton of a woman–as well as a few other bones and isolated teeth. El-Wad had only fragments of bones from at least two adults and an infant. Although Garrod had some doubts about their context and whether the skeleton might have been an intrusive burial, the bones appear to have come from the Middle Paleolithic layers (layers B and C). Another cave, Mugharet Es-Skhul, had been used for a shorter period of time, but Garrod unearthed the remains of at least ten individuals, including three skeletons. About the same time, from 1933 to 1955, a French diplomat named Rene Neuville excavated a cave at Mount Kafzeh in Galilee to the north and east of Mt. Carmel. The site, Djebel Qafzeh, had the remains of two individuals.

Arthur Keith and Theodore McCown described these human remains. Taken together they represented a highly variable population on the brink of being fully modern. The skulls mixed such primitive features as heavy brows and a large mid-face with advanced features, including a forehead, a rounded occiput (back of the skull), and a chin. However, each cranium was distinct. Keith and McCown understood these to be a common ancestral population for Neanderthals and later anatomically humans. This view was consistent with Garrod's interpretation of a continuous and gradual evolution of stone tool culture. The discoveries did much to bring Neanderthals and more primitive hominins closer to the mainstream of the human family.

Excavations in these caves and others were resumed in later decades. The number of skeletons at Qafzeh grew to 15 in the 1970s, with Skhul expanding to a minimum of 14. Kebara cave on Mt. Carmel also contained an infant. Renewed digging there in 1982 by Ofer Bar-Yosef yielded an adult skeleton remarkably complete except for the cranium and parts of the lower limbs. Hisashi Suzuki excavated at Amud, another cave north of the Sea of Galilee, in the 1960s. His team found a complete adult skeleton and parts of three others. More exploration in 1991–1992 produced three infant skeletons.

This wealth of skeletal material has permitted a better understanding of the populations and their relations to other parts of the world. The specimens from Amud, Tabun, and Kebara proved to be Neanderthals, clearly expressing the distinctive cranial and skeletal features of that people. Those from Skhul and Qafzeh represent a near modern population. The mix of primitive and modern traits that Keith and McCown described makes their exact affiliation difficult to pin down. A plausible story was not difficult to imagine: Neanderthals spread out of Europe and occupied the Near East, expanding into Iraq at Shanidar and even further east. They evolved into the later transitional "modern" people for Skhul and Qafzeh. It was still possible that the latter were a hybrid with a modern immigration from Africa, but the lack of sudden changes in culture spoke against that. All of these hominins were associated with a Mode 3 technology. All that was needed was to pin down the dates. However, when those dates became available, the story had to be rewritten.

Unexpected Dates

Putting dates on the skeletons brought together a range of disciplines and approaches, including paleoclimatic data, faunal correlations, and new techniques of absolute dating. The climatic swings that had been studied in Europe provided a framework for a relative chronology in the absence of absolute dating. Europe was believed to have four major glacial cycles, with intervening warm interglacials. The last interglacial ended about 110,000 years ago with the onset of the Würm glacial expansion. It was toward the end of this last Ice Age, when the habitability of Europe was reduced and the Neanderthal population was low, that anatomically humans made their entrance in Eastern Europe.

The Near East did not have glaciations, but there were corresponding swings in temperature and rainfall. Arthur Jelinek, who resumed excavations there in 1967, placed the sediments at Tabun in this context. The Lower Paleolithic levels (layers E and F) were sandy from a coastline that lay nearby while the sea level was high. This corresponded to the last interglacial. As the ice sheets expanded and locked up more water, the sea dropped and the shoreline shifted away from the site. Subsequently, Tabun accumulated increasing proportions of silt rather than sand in the upper layer E and layer D, which marked the beginning of the Middle Paleolithic. The local climate in Israel was wetter supported forests, as indicated by fauna and pollen. This lasted through the rest of the Middle Paleolithic (layers C and B).. Part of the roof of the cave collapsed at this time and clay washed in to mix with the fallen rocks. This crude outline placed Tabun and the cultural sequence into a relative time framework with Europe. The Neanderthals at Amud and Kebara were presumed to be roughly contemporaneous with the skeleton from level B at Tabun, while the near modern population was assumed to be present later in time.

Later work led by Ofer Bar-Yosef introduced two new techniques for absolute dating during the 1970s and 1980s. While their results are not entirely consistent, they have turned our understanding of the sequence on its head. Thermoluminescence (TL) results from the release of electrons from crystals as they are heated. Electrons become trapped in crystals over time, and counting them as they are released permits a measure of the length of time they have been accumulating. In the Mt. Carmel caves, the most datable objects are the numerous flint tools and debris. If the tools fell into the hearths and were burnt, that heating reset the "clock" and drove out the trapped electrons. By heating the objects a second time in a laboratory, scientists can calculate the length of time new electrons were accumulating since the hearths were burning.

Another relevant technique, electron spin resonance (ESR), also takes advantage of electrons trapped within crystals. Natural crystals that develop in tooth enamel and eggshell can be damaged by background radiation. The defective crystal lattice can trap electrons, and those can be detected by their behavior in a magnetic field. Assuming the background radiation is constant and known, the rate of damage to the crystals can be predicted, and number of trapped electrons will indicate age. A cave is an excellent setting to control for environmental variables such as radiation and temperature, and ESR can be applied to teeth found there.

The new dating techniques quickly confirmed the suspected dates for some sites. The Amud skeletons were 50,000–70,000 years old, based on TL with burned flints. Kebara was dated by ESR to 60,000 years. Tabun was more of a problem because repeated dates were not all consistent and because the exact stratigraphic placement of the skeleton and mandible was uncertain. The end of the Lower Paleolithic (layer E) was moved back in time to about 215,000 years ago, more consistent with European chronology. Strata C and B, where the skeleton and mandible were found, were dated by ESR to about 135,000 and 104,000 years, respectively.

There had been clues that Qafzeh was older than had been previously assumed. Two extinct species of rodents found there, *Mastomys batei* and *Avichanthis ectos*, were already familiar from older deposits in Europe. A younger species of gerbil, *Cricetulus migratorius*, had been found at Tabun in what was supposed to be a deeper stratum. These faunal signals plus climate data, including oxygen isotope ratios, were consistent with a date for Qafzeh of 80,000–100,000 years ago. Thermoluminescence confirmed a date of 92,000 years. When ESR dates from teeth at Skhul produced dates of 81,000 and 101,000, it became clear that the "modern" skeletons from these two caves actually preceded the Neanderthals from Amud and Kebara.

A Meeting of Different Continents

The new scenario raises provocative questions and new interpretations. Human presence in the Near East was continuous from the Lower Paleolithic to the present. There is direct evidence of stone tools from 1.6 Ma at 'Ubeidiya in Israel. One can assume the hominins at Dmanisi had already passed that way about 1.8 Ma. However, there is little anatomical evidence of the identity of the later inhabitants until the "modern" human skeletons at Qafzeh and Skhul around 100,000 years ago. These are not necessarily precocious—there are other samples of early populations of *H. sapiens* between 150,000 and 200,000 years ago from Herto in Ethiopia and Omo Kibish in Kenya. All three of these regional groups show considerable variation among individuals, with differing mixtures of primitive and modern traits. It is reasonable to assume that humans with modern traits evolved in Africa and then advanced north into Asia on more than one occasion (Fig. 2).

The fossils may be put in context of archaeology, fauna, and indicators of climate to attempt to reconstruct the history of populations in this region. After the time of Qafzeh and Skhul, around 50,000–70,000 years ago, Neanderthals became widespread in western Asia, as seen at Amud, Kebara, and Shanidar. If the skeleton at Tabun is correctly identified, Neanderthals may have arrived early enough to overlap with their predecessors.

Evidence concerning fauna and the environment sheds some light on these movements. As the climate oscillated, other animals moved in and out, particularly smaller mammals that are more temperature sensitive. The beginning of the Middle Paleolithic saw a cool climate with more trees. The fauna resembled animals of other parts of Eurasia that are more tolerant of cold weather. With the warmer weather about 100,000 years ago, there was an increase in species known from Africa and the Arabian Peninsula that were adapted to warmer and drier conditions. These included animals of the desert and steppes, such as ostrich, dromedary camels, and hartebeest. The return of cool weather coincided with Neanderthal presence and with Eurasian fauna again. Put another way, the African and European hominins were parts of two different continental fauna adapted to different climatic conditions. As climate changed, the boundary that separated them moved to the north or south, and the species expanded or contracted their ranges.

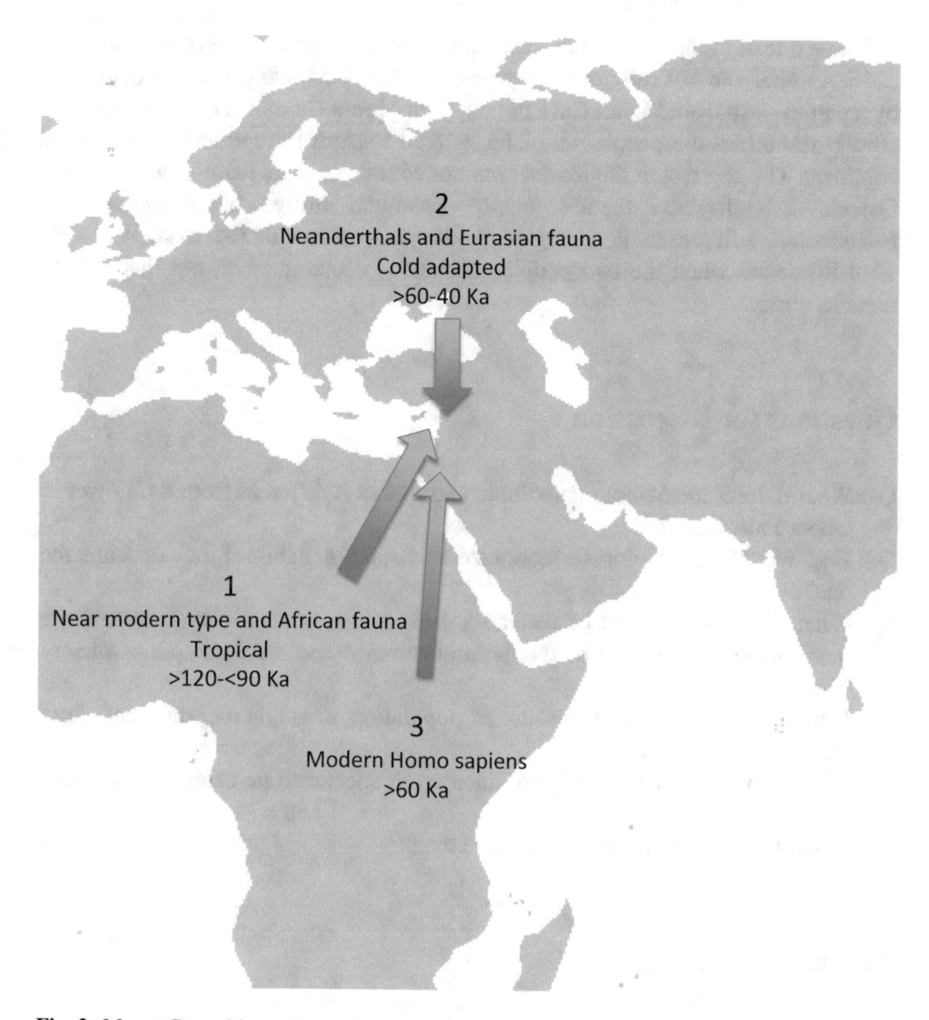

Fig. 2 Mount Carmel lay at the intersection of two biomes containing the cold-adapted Eurasian fauna and the warm-adapted African fauna. As the climate shifted, animals and humans migrated to the north or south

The "Neanderthal problem" asks for a reason why anatomically humans proved evolutionarily more successful than Neanderthals. A hundred thousand years ago, they were not any more successful. With similar cultures nearly impossible to distinguish, the two populations appear in the fossil record merely as components of their respective ecosystems playing parallel roles in different local faunal communities. Each proved vulnerable to changes in the habitat to which it was adapted.

These questions have been raised again by a most recent discovery announced in 2015. Another wave of fully modern people arrived by 55,000 years ago, evidenced by a partial skull from Manot Cave not far from Mount Carmel. This cranium more closely resembles the people who later entered Eastern Europe and occupied the continent. The contrast of the Manot cranium with the archaic features at Skhul and Qafzeh likely displaces them to another dead-end lineage, but it overlaps the Neanderthals still present in the region. Anthropologists would like to know exactly what happened when the two peoples encountered one another, but that answer remains elusive.

Questions for Discussion

Q1: What defines the Middle Paleolithic and makes it different from the Lower or Upper Paleolithic?
Q2: How would paleontologists recognize a hybrid population if they encountered one?
Q3: What limitations (types of materials, time range, contaminating factors) are there on the application of thermoluminescence and electron spin resonance dating?
Q4: Is there such a thing as a transitional population, or is that a by-product of our arbitrary classifications?
Q5: Is it reasonable for two different lineages or species to be using identical tool cultures?
Q6: What defines anatomically modern humans?

Additional Reading

Bar-Yosef O, Vandermeersch B (1993) Modern humans in the Levant. Sci Am 268(4):94–100
Hershkovitz I et al (2015) Levantine cranium from Manot Cave (Israel) foreshadows the first European anatomically humans. Nature 520:216–219
Jelinek AJ (1982) The Tabun Cave and Paleolithic man in the Levant. Science 216:1369–1375
Shea JJ (2001) The Middle Paleolithic: early modern humans and Neandertals in the Levant. Near East Archaeol 64:38–64
Shea JJ (2003) Neandertals, competition, and the origin of modern human behavior in the Levant. Evol Anthropol 12(4):173–187
Valladas H et al (1988) Thermoluminescence dating of Mousterian 'Proto-Cro-Magnon' remains from Israel and the origin of modern man. Nature 331:614–616

Case Study 21. Chasing Smaller Game: The Archaeology of Modernity

Abstract In the well-studied ancient landscape of Europe, there was a sudden change starting around 46,000 years ago, as modern people replaced Neanderthals and the Middle Paleolithic gave way to the Upper Paleolithic. The cultural differences are dramatic, and it is difficult to avoid the conclusion that this represents first appearance of modern human minds as well as bodies. The search for where and when they developed leads to Africa. Defining modern behavior is difficult, however, and depends on interpretations of indirect evidence. One realm where there is more direct evidence of behavior is subsistence and diet. Thc cultures of anatomically modern people are characterized by exploitation of an increased diversity of prey and other food resources. While this trend indicates expanding technological abilities, it may also reflect the ecological necessity to support an expanding population. Population pressure and more frequent encounters between groups are likely both contributors to and consequences of the modern mind.

The Upper Paleolithic in Europe witnessed the rapid appearance of many innovations that increased the efficiency of work and signaled more complex minds. This is especially striking after the slow pace of change in the preceding two and a half million years of material culture. Although the boundary between Middle and Upper Paleolithic is briefly blurred, the contrasts before and after the transition period are marked. Later Europeans introduced blades, more precise techniques for making stone tools, the use of new materials such as bone and antler for tools, beads and other personal ornaments, representational art, needles, bows and arrows, atlatls (spear-throwers), rope, barbed harpoons, and musical instruments. The advances were so dramatic that some anthropologists suspect a much more revolutionary innovation lay behind them all—the modern brain (Fig. 1).

Most of these creations, excepting representational art, have now been found in older sites in Africa. As Sally McBrearty and Alison Brooks have argued, there, the "revolution" was really a gradual transition. Some of the finds associated with the European Upper Paleolithic after 40,000 years ago, such as blades and the use of natural mineral pigments, occur much deeper into the past in Africa. Shell beads are

J.H. Langdon, *The Science of Human Evolution*,
DOI 10.1007/978-3-319-41585-7_21

Fig. 1 Blades, bone tools, and engraved ochre from Blombos Caves. Copyright Chris Henshilwood, GNU Free Documentation License with permission

reported from 135,000 years ago at Skhul Cave and 82,000 years ago at Grotte des Pigeons in Morocco. Thus, innovation began much earlier in Africa. It proceeded slowly and then began accelerating in the last 60,000 years (Fig. 2).

One site in particular, Blombos Cave on the eastern coast of South Africa, exemplifies this. Deposits there extend back 100,000 years. For most of that time, a Middle Stone Age (MSA) technology predominates. The MSA is roughly a chronological and technological counterpart to the Mousterian Culture in Europe, to be succeeded by the Late Stone Age or LSA. However, the MSA levels at Blombos also include bone points and perforated shells. A single bone fragment has engraved lines of unknown meaning. A piece of red ochre has a pattern of lines etched into one face of it. This might be the oldest symbolic or notational engraving yet discovered. Later, an artist's tool kit, including materials for mixing pigments, was discovered in this cave (Fig. 1).

Changing Subsistence Patterns

The appearance of nonutilitarian artifacts allows speculation about how Middle Paleolithic people may have been thinking. Evidence of subsistence activities demonstrates what they were actually doing. Richard Klein has conducted a series of

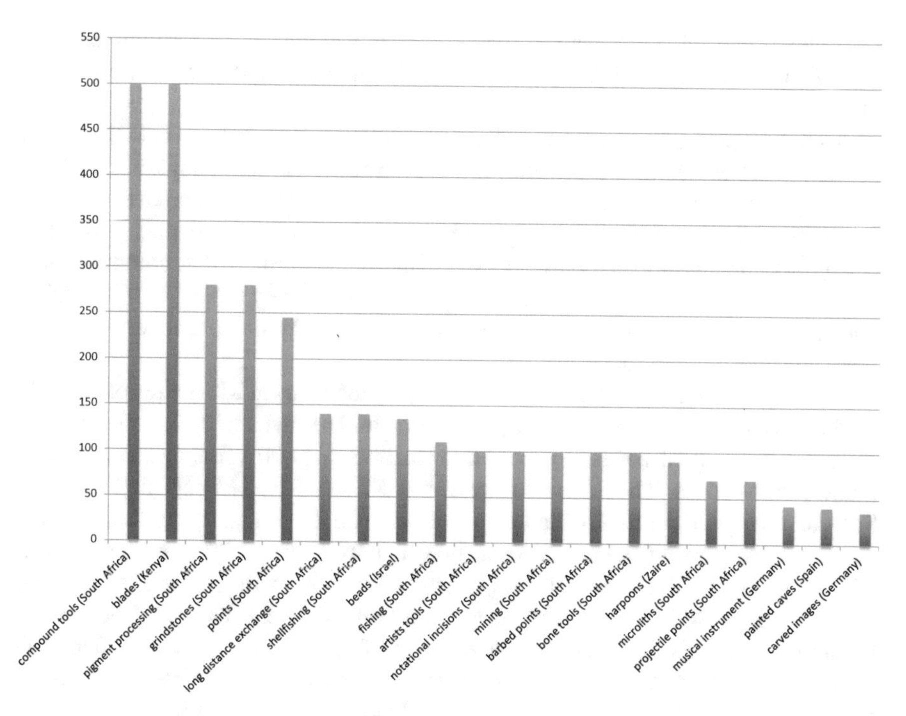

Fig. 2 First appearance of traits commonly associated with modern human behavior

close examinations of the animal bones accumulated by early humans in MSA and LSA sites in South Africa. His analysis of frequencies and types of species led him to conclude that the two periods witnessed different styles of hunting.

Klein used age classes of animals to distinguish two types of mortality patterns, catastrophic and attritional. The normal distribution of ages in an animal population is a declining curve, with more infants than juveniles because some of the infants have died, more juveniles than adults, and so on, assuming the age classes cover comparable age spans. If a herd of such animals were suddenly killed by a flash flood, the bodies would reflect that distribution. Klein refers to this as a catastrophic mortality profile. Commonly, the most vulnerable individuals, the very young and very old, experience the highest mortality rates due to predation, disease, and debility. If only the carcasses of those taken by disease or predation were examined, there would be higher frequencies at the extremes of age and very few in the prime of life. This is an attritional mortality profile.

Klein assessed age in ungulates by the height of the molar tooth crowns on the assumption that teeth wear uniformly across the age span, and by the presence of deciduous teeth (indicating very young animals) or third molars (from fully mature animals). At the MSA sites of Klasies River Mouth and Die Kelders, the most common ungulate was eland. These are large, relatively docile herd animals that might represent a preferred prey species. The age profiles for eland remains fit the catastrophic mortality pattern. A less common ungulate was the buffalo, which is more dangerous to hunt and which showed an attritional mortality pattern. At the

LSA site of Nelson Bay Cave and others, buffalo were far more common and also showed attritional mortality, while eland were more heavily skewed toward young individuals.

Klein interpreted these data as indicating a difference in abilities between MSA and LSA hunters. MSA people concentrated on relatively less dangerous animals and a style of hunting that resulted in mass kills, such as driving them over a cliff or into a trap. LSA hunters were more likely to hunt buffalo and pigs and other dangerous animals because they had weapons such as arrows that could kill at a distance. Like predators of all species, they were more likely to take the vulnerable young and old individuals. He postulated that the "revolution" of modernity was a rapid advance in brain function.

Further evidence consistent with this model is the rarity at MSA sites of animals harder to catch, such as birds and aquatic animals. The first shellfish, which can be collected in the tidal zones, are now known from 164,000 years ago at Pinnacle Point, South Africa. They are also present at Klasies River Mouth slightly later. Seals and penguins, both of which may be hunted on the beach, are present, but not common. In contrast, fish bones do not show up until after 50,000 years ago, but are present in relative abundance at LSA coastal sites along with fishing gear, such as sinkers and "hooks" which were made from bone splinters that could be baited. Flying birds are also present in LSA deposits. The presence of many infant seal remains indicates that LSA hunters were exploiting the seasonal birthing practices of seal colonies. MSA foragers appear not to have capitalized on this resource.

There is a similar pattern observed in Europe. Mousterian sites rarely show evidence of fishing. (One notable exception is a Neanderthal site at Gibraltar where marine resources were exploited.) Animal remains at a given site are likely to be dominated by a single prey species, such as reindeer or horse. Later Upper Paleolithic sites exploit a greater diversity of animals, including fish.

This reading of the faunal remains is supported by isotope analysis of hominin bones. Ratios of stable isotopes of carbon and nitrogen (^{13}C: ^{15}N) vary in different food categories, especially between fish and mammals. A series of studies by Michael Richards and Herve Bocherens have analyzed the protein content in the bones of Neanderthals and early modern Europeans and of Pleistocene game animals and attempted to match those with proportions of different foods in the diets. Neanderthals from Croatia, France, and Belgium apparently had a diet that consisted predominantly of meat, including both larger mammals—mammoth and rhinoceros—as well as smaller mammals—such as reindeer and horses. Upper Paleolithic skeletons show more diversity, including a fish component in most samples, with the addition of marine mammals in Britain.

Changing Resource Bases

Curtis Marean presents a different interpretation of Klein's data. He sees the most important trend from MSA or Middle Paleolithic to modern cultures as one of increasing dietary breadth. Efficient hunters are able to concentrate on a prey species that is preferred either because it is easy to catch or because it is more abundant, or both. While Klein believed the distinction is based on ease of capture by mediocre hunters, Marean suggested LSA populations were broadening their base of subsistence because preferred prey was becoming harder to find. Buffalo and pigs are still dangerous for hunters with arrows and javelins because they must be attacked within throwing or shooting distance. Fish and birds require specialized technology, but even with that technology they yield a much lower return for effort than a large ungulate. LSA and Upper Paleolithic hunters appear to have been compelled to exploit whatever animal resources they might encounter.

This argument is supported by the work of J. Tyler Faith. Examining a broader number of MSA and LSA sites, Faith found little evidence that the earlier cultures were less capable or more restricted in their exploitation of ungulates. MSA sites had the same size range and diversity of species after accounting for environmental differences. The tendency for large ungulates, including eland, to be more highly represented in the MSA sites was again better explained by higher prey populations or encounter rates.

Why did prey populations decline in the LSA and Upper Paleolithic? Both Klein and Faith attempted to factor in the fluctuating climate by comparing sites from interglacial periods when the environments should have been similar. However, another widely recognized difference was a larger human population size in the later period. This is apparent from the increased number of archaeological sites. Higher populations risk overhunting and driving prey away.

The effects of overhunting have been recorded by Klein and others. Klein observed an absolute decrease in size of tortoises, Cape turban shells, and limpets between MSA and LSA sites in South Africa. Both are continuously growing species whose size reflects age. Similar observations were made by Mary Stiner for shellfish and tortoises from Israel and Italy on the Mediterranean coast. The same people began exploiting hares and flying birds—prey smaller and more difficult to capture—just as had occurred in South Africa. The standard explanation for a size reduction is that intensive harvesting reduces the average lifespan of prey and the probability that a given individual will grow into the larger size category. The same phenomenon is reported for fish as they are overharvested in the twenty-first century.

Explaining the Transition

Klein argued that MSA hunters were limited in their effectiveness perhaps because they were limited in brain capacity; thus they concentrated on larger, easier to capture prey. As brain evolution crossed a critical threshold, LSA and later hunters were more proficient at exploiting a great diversity of resources. This supported a larger population, which in turn put pressure on the resources. Marean's response was that MSA hunters were equally proficient, but because there were fewer of them, they were able to focus on prey that had more return for less effort. Later hunters were forced to work harder to support the growing population. In other words, the success of MSA hunters drove them to become LSA hunters.

The difference between these two arguments is subtle and challenges our understanding of what limits population size. If that limitation is food supply, then the rate of population growth should reflect foraging proficiency. If both populations were equally adept, why weren't MSA populations as large as later ones? Human populations were probably not held back by the climate fluctuations alone. The previous interglacial was not shorter than the current one and would have provided plenty of time for the population to respond. The Upper Paleolithic population expansion began under Ice Age conditions in Europe as Neanderthals were declining and continued through the last glacial expansion. The phenomenal human potential for population growth under ideal conditions has been demonstrated repeatedly in recorded history. Some other factor must have been involved.

For any other large species, it is assumed that the population is in rough equilibrium with carrying capacity. Carrying capacity is a theoretical concept in which food supply and other limiting resources determine the maximum sustainable population. MSA humans presumably also operated at their carrying capacity until that carrying capacity changed. In the absence of evidence for changes in human anatomy during this period, culture, not brain evolution, is the most parsimonious explanation. Language, blades, symbolic thought, compound tools, better weapons—any of these could increase the ability to extract food by incremental amounts.

What explains this sudden cultural change? Klein and others suggest a rapid biological evolution of the brain, but such speculation is probably unnecessary. In the preceding 2 Ma of human history, people lived in small isolated hunter-gatherer bands. Technology did not visibly change for hundreds of thousands of years at a time. Only the most exceptional innovations by themselves could have made a significant difference in survival. Thus, it is easy to imagine that true advances were quite rare, and, when they did occur, they were quickly lost with the death of the individual or extinction of the band. What gradually changed were population density and the opportunity to exchange ideas. As the frequency and complexity of social interactions between groups increased, the ability to share, preserve, and combine ideas spurred meaningful cumulative innovation. Innovation opened new opportunities; thus, a positive feedback would become possible.

The history of human technology may be read as a quest to employ culture to expand carrying capacity. The "modern revolution" began in Africa and progressed slowly at first, then with accelerating pace in the past 60,000 years when a critical threshold of population was achieved. In Europe, after a slow start roughly equal to the amount of time it took humans to spread across the continent, cultural evolution proceeded explosively. The population grew as a consequence and subsistence activities were altered to support more people on less land. The Upper Paleolithic was succeeded in Europe by the Mesolithic, which is most notable for regional cultural diversification and more intensive and ingenious exploitation of resources. Population pressure continued throughout the Old World, eventually requiring people to find even more effective resource extraction strategies: animal domestication and agriculture.

Questions for Discussion

Q1: McBrearty and Brooks argue that the innovations of modern humans were evidence of mental capacities for abstract thinking, ability to plan, and symbolic behavior. What specific traces in the archaeological record would enable us to recognize these traits?

Q2: If one society lives off eland and another rabbits, which group is better at hunting? What information is needed in order to address this question?

Q3: Klein and Marean present alternative hypotheses to explain the data. How these be tested? Why might one hypothesis be considered better than another?

Q4: Does the fact that Neanderthals could produce Upper Paleolithic artifacts in the Chatelperronian culture, but only after contact with newly arrived culture, tell us anything more about Neanderthal intellectual capacity?

Q5: We speak of other revolutions in technology involving the introduction of agriculture, metals, mechanized travel, or computers. Are these qualitatively similar to or different from that of the Upper Paleolithic?

Additional Reading

Bocherens H et al (2001) New isotopic evidence for dietary habits of Neanderthals from Belgium. J Hum Evol 40:497–505

Faith JT (2008) Eland, buffalo, and wild pigs: were Middle Stone Age humans ineffective hunters? J Hum Evol 55:24–36

Henshilwood CS et al (2002) Emergence of modern human behavior: Middle Stone Age engravings from South Africa. Science 295:1278–1280

Klein RG (1983) The Stone Age prehistory of southern Africa. Annu Rev Anthropol 12:25–48

Klein RG (2009) The human career: human biological and cultural origins, 2nd edn. University Chicago Press, Chicago

Marean CW (2010) When the sea saved humanity. Sci Am 303(2):55–61

Marean CW, Assefa Z (1999) Zooarcheological evidence for the exploitation behavior of Neanderthals and early modern humans. Evol Anthropol 8(1):22–37

Marean CW et al (2007) Early use of marine resources and pigment in South Africa during the Middle Pleistocene. Nature 449:905–908

McBrearty S, Brooks AS (2000) The revolution that wasn't: a new interpretation of the origin of modern human behavior. J Hum Evol 39:453–563

Pringle H (2013) The origins of creativity. Sci Am 308(3):36–43

Richards MP et al (2000) Neanderthal diet at Vindja and Neanderthal predation: the evidence from stable isotopes. Proc Natl Acad Sci U S A 97:7663–7666

Richards MP et al (2001) Stable isotope evidence for increasing dietary breadth in the European mid-Upper Paleolithic. Proc Natl Acad Sci U S A 98(11):6528–6532

Steele TE, Klein RG (2007) Late Pleistocene subsistence strategies and resource identification in Africa. In: Hublin J-J, Richards MP (eds) The evolution of hominin diets: Integrating approaches to the study of Paleolithic subsistence. Springer, New York, pp 113–126

Stiner MC et al (1999) Paleolithic population growth pulses evidenced by small animal exploitation. Science 283:190–194

Stiner MC et al (2000) The tortoise and the hare: small-game use, the broad-spectrum revolution, and Paleolithic demography. Curr Anthropol 41:39–73
Stringer CB et al (2008) Neanderthal exploitation of marine mammals in Gibraltar. Proc Natl Acad Sci U S A 105(38):14319–14324
Wong K (2005) The morning of the modern mind. Sci Am 292(6):86–95
Wynn T, Coolidge FL (2008) A Stone-Age meeting of the minds. Am Sci 96:44–51

Case Study 22. The Cutting Edge of Science: Kissing Cousins Revealed Through Ancient DNA

Abstract In Michael Crichton's science fiction novel Jurassic Park, scientists recover the blood of dinosaurs and use their DNA to bring *Tyrannosaurus rex* back to life. Bringing dinosaurs to life is still science fiction, though a number of laboratories are working on mammoths and some other recently extinct mammals for which soft tissues are available. However, it is now possible to recover and sequence DNA from fossils that have been preserved under the proper conditions. As genome sequencing becomes faster and cheaper, we are able to ask questions of the fossils that previously were only encountered in fiction. There is every reason to believe that these studies are in their infancy, but they are already helping us to reimagine the prehistoric landscape.

Recovering Ancient DNA

The development of tools for sequencing and analyzing the genome very quickly gave rise to dreams of recovering DNA from extinct animals. That this is even conceivable for older animals is because we can recover DNA from bones. A scientist wanting to sequence ancient DNA faces two major hurdles. The first is partial degradation, in which the surviving DNA has broken into small segments. The presence of water and heat speed the process, as do bacteria feeding off the dying tissues. Imagine trying to reconstruct a building that has been taken apart. If the pieces are too small, the task is hopeless. This problem can be overcome, however, because the fossil contains many copies of the DNA that are broken in different ways. If the segments are long enough, corresponding sections can be matched together where they overlap to create still longer ones. The problem is a little easier if we know what the original structure looked like. The close similarities between human DNA and that of other hominins help us to identify the location and function of genes even if they have been slightly altered.

The second problem is contamination. Bacteria, molds, and environmental detritus can introduce fragments of DNA, and so can anthropologists handling the bones in the field or the laboratory, or even walking through the laboratory where they are

J.H. Langdon, *The Science of Human Evolution*,
DOI 10.1007/978-3-319-41585-7_22

housed. Any technique to extract and copy the original DNA will act on the DNA from all of these sources. Chimpanzee DNA is approximately 97 % identical to that of a human, and we can be certain that any hominin DNA is even closer. Therefore, if DNA is being extracted from a fossil human, it is usually possible to distinguish that from contaminating bacteria or mold. If the DNA sequence that was extracted does not look similar to human, it must be contamination. Unfortunately, people in the museum and the laboratory are a likely source of extraneous DNA, so if the extracted sequence appears similar to humans it may still be contamination.

The German geneticist Svante Pääbo found ways to overcome contamination and piece together the genome of human ancestors, including Neanderthals. After numerous failures, strict protocols have been worked out to maintain a clean laboratory that minimizes the presence of bacteria or the tendency of humans to shed their own cells. Specimens are only handled with clean gloves, and attempts are made to recover DNA from the center, not from the surface of the bones and teeth. It is also practice to maintain a profile of everyone potentially in contact with the lab and specimens, so that modern human contamination can be identified. Pääbo's work has led to new understanding about past relationships among different species of humans and hint at the existence of populations not currently known from the fossil record.

Neanderthal Genes

Svante Pääbo was the first to obtain a partial sequence of mitochondrial DNA from an extinct hominin in 1997. He looked for mtDNA first, because it is much more common than nuclear DNA, with potentially thousands of mitochondria in a single cell. He chose, appropriately, the type specimen of Neanderthal from the Feldhofer Quarry in Germany, which was estimated to be 30,000–100,000 years old. The resulting sequence was sufficiently distant from modern humans to suggest a long separate history for our two species with a divergence time between 550 and 690 Ka. It seemed to negate the long-debated possibility that Neanderthals were our ancestors.

MtDNA from a second Neanderthal, this time from Mezmaiskaya Cave in the Caucasus, was sequenced in 2000 and proved similar, but not identical to the first. Within a decade sequences, including complete mtDNA, were available for at least 15 individuals from Germany, Russia, France, Spain, Croatia, Belgium, and Italy, and the number continues to grow. Most of them date from near the end of the Neanderthal era, less than 50,000 years old, but one was nearly 100,000 years old. With a number of individuals sampled, it became possible to ask different questions. All of the fossils showed similarities, indicating they came from the same matrilineage distinct from that of modern humans. However, not unexpectedly, there were differences among the Neanderthals. At last three clusters are apparent that correspond to Western and Southern Europe and the Middle East.

Pääbo then tackled the nuclear DNA, announcing an outline of the Neanderthal genome in 2006, with increasingly complete sequencing in subsequent years. The Neanderthal genome proved to be 99.5 % identical as modern humans and equally distinct from the Upper Paleolithic peoples that replaced them in Europe. A common ancestor with us was calculated to have existed between 390,000 and 500,000 years ago.

In order to learn more about the Neanderthals themselves, the researchers began by looking for genes of known functions in humans. They found, for example, that at least some Neanderthals had pale skin and red hair. Interestingly, they also carried a modern variant of the gene FOXp2 that some researchers had suggested corresponded to modern speech.

The issue of Neanderthal language had eluded researchers for a century, residing more in imagination than in evidence. Many researchers had attempted to find indicators in the skeleton that would tell us whether Neanderthals could talk like we can. For example, as early as 1971 Philip Lieberman attempted to use the base of a Neanderthal cranium to reconstruct the pharynx and identify the range of sounds a Neanderthal could make. Other studies focused on the shape of the hyoid bone or the size of the canal through which the hypoglossal nerve passes on its way to the tongue. Although this work generally failed to find convincing functional differences from humans, all of the indicators were controversial and inconclusive.

Where anatomical studies failed, genetics offered a different approach. In 2001 clinicians identified a family with heritable language disorders that could be traced to the FOXp2 gene. This codes for a membrane protein expressed in neurons. Its relationship to language is not understood, but it may facilitate certain patterns of neural communication. Studies by Pääbo's laboratory reported that the normal sequence of the gene in modern humans is unique among living primates and acquired two mutations quite recently, probably within the past 100,000 years. If the new allele could become fixed in the human population so rapidly, it must have been under strong positive selection. The date appeared to coincide with the migration of modern humans out of Africa (see Case Study 16) and fit with models suggesting some extraordinary advantage enabled them to displace archaic populations in Europe and Asia. While at first glance the discovery of the modern form of the FOXp2 gene in Neanderthals suggests they had language, that same find undermines the previous work. Neanderthals and therefore our immediate ancestors likely acquired it well before 1000,000, and the trait could not explain a competitive difference.

The genetic differences between Neanderthals and modern populations are distinct, but quite small. They have not led us to any greater understanding of why Neanderthals are extinct and why we are the only species alive today. The possibility remains that we may yet discover explanations for the skeletal peculiarities of Neanderthals and other archaic forms of humans. It is likely that such important shifts in function reflect modifications to controls that up-regulate or down-regulate metabolic pathways in the brain or other tissues rather than changes in structural genes.

Overall the sequences from different Neanderthal individuals that have been observed show a low degree of genetic variability, consistent with the hypothesis that the Neanderthal population in glacial Europe was never very large. One sample from Denisova Cave, Siberia, in the extreme eastern edge of the Neanderthal range shows evidence of extensive inbreeding, probably because of isolation. Examination of the mtDNA of 12 individuals at El Sidron in Spain shows that normally females moved between social groups. The three adult males in this cave shared the same mtDNA, but the three females and six children did not.

Denisovan Genes

In 2008 a fragment of a 40,000-year-old finger bone was found in Denisova Cave in central Siberia. Such a small fossil had little to tell us anatomically, but it yielded some interesting DNA. Remarkably, it represents a population of humans as distinct from Neanderthals as both are from modern humans. Because the only other remains associated with the finger bone were two teeth, we cannot associate these bones and this genome with fossils from other localities or from any known type of human. They are being referred to simply as the Denisovans. Nonetheless, we now know that, along with *Homo floresiensis*, there were at least four and most likely more species of humans alive at this time. As modern humans spread across the Old World, they interbred to some extent with the archaic populations they encountered.

A clue to the origin of the Denisovans came with the sequencing of a much older individual. Sima de los Huesos in northern Spain is one of the richest sites in the world for premodern hominins. This population appears to be transitional between *H. heidelbergensis* and *H. neanderthalensis*, more than 300,000 years old. In 2014, Pääbo's lab published the mtDNA sequence from one of the bones found there, the oldest mtDNA recovered and sequenced to date. The mtDNA more closely resembled that from Denisova than it did Neanderthals. The simplest explanation is that a single population gave rise to both Neanderthals and Denisovans, or at least to the two corresponding matrilineages. Since none of the later Neanderthal samples match the Denisovan sample, current evidence suggests the latter genome type became rare in the west but persisted in Asia.

The Fate of Neanderthals and Other Archaic Humans

The revelations from ancient DNA now make it possible to address the Neanderthal problem from a new angle. Did the Neanderthal genome disappear entirely? No. Modern Eurasians share some genes with Neanderthals that are not seen in modern Africans. Neanderthal genes make up to 1–4 % of the genome of different individuals and must have entered the modern population through limited interbreeding, although we now know that there were a number of hybridization events.

There were many opportunities for the two populations to mingle. We know that moderns appeared in Israel 60,000 years ago while Neanderthals were still living there. A cranium and mandible from Oase Cave in Rumania about 40,000 years ago show Neanderthal traits on a fundamentally modern morphology. Chunks of Neanderthal DNA sequence in the genome of the mandible indicate interbreeding had occurred within a half dozen generations. In western France, changes in the archaeological remains associated with Neanderthals suggest cultural changes, which would have provided opportunities for cross-breeding. In Spain and the Caucasus the last Neanderthals persisted in refuges where they might have overlapped in time with moderns. It would be most surprising if the two populations did not interbreed in many places.

Given these observations, it would also be surprising if modern genes had not entered the Neanderthal population. Interaction at some level had long been suspected on the basis of cultural innovations among late surviving Neanderthals and from a few skeletal finds that mixed archaic cranial features with more modern postcrania. A 2016 study identified modern genes in the Neanderthal bone from Denisova that were not present in Western European Neanderthals. Comparison with modern humans shows this genetic material is most closely related to that found in African populations and that it likely entered the Neanderthal population before 68,000 years ago. This indicates contact between the two peoples occurred significantly earlier than the time modern fossils have appeared in Western Asia and Europe. Possibly there was an earlier migration out of Africa that left no descendants or the main migration did not spread as rapidly as has been assumed.

Denisovan DNA can also be detected among modern populations. It comprises 1–6 % of the genome of people of Australia, New Guinea, and Melanesia and in lesser amounts in India and East Asia. Again, there is evidence for several separate interbreeding events. There are many reasons, however, why such a small percentages of archaic DNA survives today. It is possible that selection acted against most Neanderthal genes. It is also likely that any specific societies on the frontier of expansion by the moderns went extinct long ago and were replaced by later migrations; thus more highly hybridized populations may have disappeared by chance. However, the borrowed genes that persisted in the modern genome may have been kept for a reason.

It is possible now to scan the modern human genome for these borrowed genes and to use the comparisons between our species to explore the meaning of the differences. While exact functions of specific genes are generally difficult to understand at this time, we have some clues according to the tissues where they are expressed. We know, for example, Neanderthals had a number of genetic changes involved in bone development and the skin. In contrast, the modern genome had undergone more change relating to pigmentation. There are very few Neanderthal genes on our X chromosome and none on the Y and the modern genome has more unique genes expressed in the testis. All of this suggests heavy selection against potential fertility problems caused by hybridization.

A number of genes inherited from Neanderthals have medical implications, including a greater risk of heart attack because of rapid blood clotting and increased risk of depression, sun sensitivity, and susceptibility to nicotine. On the other hand, three genes have been identified that boost our immune defenses. The latter genes probably were selected for, and the others may be connected with some subtle benefits. One functional gene inherited from the Denisovans helps people adapt to high altitude. Ironically, while this might have been adaptive in the Altai Mountains of Siberia where the Denisovan finger bone was found, it is less useful in Southeast Asia or Pacific Islands where Denisovan genes are more common today.

We have learned from the Human Genome Project that a listing of human genes is not sufficient to understand how our chromosomes determine human structure and function. Chemical modification of the DNA strands, including the attachment of methyl groups, can regulate or silence individual genes. These epigenetic changes are critical for cell differentiation and normal development patterns. Methylation affects the way DNA degrades over time and thus leaves a signature in ancient DNA. Geneticists were able to recover some patterns from the Neanderthal genome. In particular, there was excess methylation near HOX genes involved in limb development. It was suggested that this might explain Neanderthals' proportionately shorter arms and legs.

Beyond Ancient DNA

The discovery of Denisovans was a complete surprise, but perhaps it should not have been. Eurasia has been populated, albeit sparsely, for nearly 2 Ma. Current models understand that modern humans emerged from Africa within the past hundred thousand and Neanderthals were restricted to Europe and Western Asia. We should have been asking who occupied the rest of the hemisphere. The fossil record for Africa, pre-Neanderthal Europe, and Eastern Asia is complex and fossils refuse to be easily sorted into lineages. Multiple morphological groups appear to coexist in time within each of these regions. Moreover, most of South Asia and Southeast Asia as well as West Africa have no fossil record, but the presence of stone tools tells us these were not uninhabited. Many of these peoples must have interbred with our ancestors. If we are able to recover more ancient DNA, we will certainly discover populations that are currently unknown. In the meantime, we can scan our own DNA for evidence of past interbreeding.

There is evidence within the Denisovan genome of an introgression of genes from still another unknown people. Some of these genes can be found in modern East Asian peoples. A similar discovery was made in 2011 by Michael Hammer in a 13,000-year-old cranium from Iwo Eleru in Nigeria. While considered fully modern by age, this skull still has a primitive overall elongated shape consistent with its mixed genetic heritage. There is no evidence that hybridization event left its mark on people today; however, another introgression was later found by Hammer in a Central African Pygmy population. The Pygmies, along with the Bushmen of South Africa, are peoples native to African long assumed on the basis of unique language distinct body statute to have had a long separate history and partial isolation from their neighbors. Hammer's team estimates an admixture with an unidentified population as recently as 30,000 years ago.

Odd hominins were found in Red Deer Cave (Maludong) in China in 1989 as recent as 12,000 years old. They are considered modern humans and not primitive, but anatomically unique. *Homo floresiensis* existed until only 13,000 years ago. The most recently named species, *Homo naledi* is undated. While its morphology and small brain size suggest a separate lineage as far back as the earliest Pleistocene, the evidence for deliberate disposal of the dead in Rising Star Cave in South Africa would not be expected until the Late Pleistocene. Anthropologists have been most reluctant to abandon the nineteenth-century paradigm that describes human evolution as a chain of species ascending to *Homo sapiens*, but evidence is accumulating that the story is far more interesting than Haeckel could have imagined. With the help of new genetic tools we will be able to see the fossil record in a different light and likely find cousins we never suspected we had.

Questions for Discussion

Q1: How would you expect the presence of multiple species of coexisting hominins to appear in the fossil and archaeological records? Is that what we observe?

Q2: DNA from archaic hominin lineages is responsible for only small proportions of the modern genome. What reasons might explain this?

Q3: If multiple hominin species existed at any one time until recently, why is there only one species now?

Q4: What other kinds of information might we hope to obtain by studying ancient DNA?

Additional Reading

Bustamante CD, Henn BM (2010) Shadows of early migrations. Nature 468:1044–1045

Callaway E (2011) Ancient DNA reveals secrets of human history. Nature 476:136–137

Castellano S et al (2014) Patterns of coding variation in the complete exomes of three Neandertals. Proc Natl Acad Sci U S A 111:6666–6671

Fabre V et al (2009) Genetic evidence of geographical groups among Neanderthals. PLoS One 4(4):e5151

Gibbons A (2011) African data bolster new view of modern human origins. Science 334:167

Gibbons A (2016a) Neandertal genes linked to modern diseases. Science 351:648–649

Gibbons A (2016b) Five matings for modern Neandertals. Science 351:1250–1251

Hammer MF (2013) Human hybrids. Sci Am 308(5):66–71

Hawks J (2013) Significance of Neandertal and Denisovan genomes in human evolution. Annu Rev Anthropol 42:433–449

Holden C (1998) No last word on language origins. Science 282:1455–1458

Hsieh PH et al (2016) Model-based analyses of whole-genome data reveal a complex evolutionary history involving archaic introgression in Central African Pygmies. Genome Res 26:291–300

Krause J et al (2007) The derived FOXP2 variant of modern humans was shared with Neandertals. Curr Biol 17:1908–1912

Krings M et al (1997) Neandertal DNA sequence and the origin of modern humans. Cell 90:19–30

Kuhlwilm M et al (2016) Ancient gene flow from early modern humans into Eastern Neanderthals. Nature 530:429–433

Lalueza-Fox C et al (2011) Genetic evidence for patrilocal mating behavior among Neandertal groups. Proc Natl Acad Sci U S A 108:250–253

Meyer M et al (2014) A mitochondrial genome sequence of a hominin from Sima de los Huesos. Nature 505:403–406

Pääbo S (2014) Neanderthal man: in search of lost genomes. Basic Books, New York

Pennisi E (2014) Ancient DNA holds clues to gene activity in extinct humans. Science 344:245–246

Reich D et al (2010) Genetic history of an archaic hominin group from Denisova Cave in Siberia. Nature 468:1053–1060

Sankararaman S et al (2014) The genomic landscape of Neanderthal ancestry in present-day humans. Nature 507:354–357

Skogland P, Jakobsson M (2011) Archaic human ancestry in East Asia. Proc Natl Acad Sci U S A 108:18301–18306

Case Study 23. Is Humanity Sustainable? Tracking the Source of our Ecological Uniqueness

Abstract Behind all of our attempts to understand human evolution is our curiosity about ourselves. "Human nature," the innate drives and desires that define us as a species, has always been constructed according to our own prejudices and aspirations, but it is rarely possible to investigate such models through science. A common approach has been to compare ourselves to other animal species for clues of what lies beneath our cultural veneer. Choosing the appropriate animal model is not less subjective. In this study, humans are compared to other mammals seeking ecological similarities and differences to clarify which might be appropriate models for us as we ask the question, "When did humans become unique?"

In 2003, marine ecologists Charles W. Fowler and Larry Hobbs asked the question, "Is humanity sustainable?," contemplating the immediate economic and ecological issues of modern human society. Specifically, Fowler and Hobbs tested the hypothesis that "the human species falls within the normal range of natural variation observed among species for a variety of ecologically relevant measures." By comparing humans with a wide range of terrestrial and marine mammals and birds, they demonstrated that humans are outliers with respect to many ecological parameters, including CO_2 production, energy and biomass consumption, geographical range size, and population size. Fowler and Hobbs rejected their hypothesis and observed that by behaving outside the range of normal parameters we disrupt the equilibrium of our ecosystem in ways that are not sustainable.

One of the core premises of the discipline is that paleoanthropologists view humans as another species of primate, subject to the same biological principles as any other species. When in the course of human evolution did such abnormality arise? Is it a property of the species or only a consequence of a complex agricultural or industrial society? While Fowler and Hobbs evaluated industrial society, this chapter compares samples of modern and prehistoric hunter-gatherers with other medium-sized mammals. It looks at life history and other ecological parameters for 109 living genera of medium-sized placental mammals. Humans are represented by prehistoric and hunter-gatherer populations.

J.H. Langdon, *The Science of Human Evolution*,
DOI 10.1007/978-3-319-41585-7_23

Life History Strategies

Life history strategies concern the allocation of energy to development, maintenance, and reproduction. Species may choose to mature and reproduce earlier, skimping on the opportunity to grow larger and store more resources or delay reproduction until later in life, using the additional years to become bigger, stronger, more competitive, or more intelligent. It is not surprising, therefore, that body size correlates with a longer lifespan. Large brains demand a great diversion of resources and shape themselves through early experience; thus, brain size also correlates with longevity. From body and brain size, it is possible to calculate an expected rate of development.

Having relatively large brains for their body size, the great apes develop slowly and live a long time compared to other mammals. Humans take this trend to an extreme degree; however, in terms of longevity and age of maturation, we are most like the apes (Figs. 1 and 2). In one parameter, however, humans depart from the depicted values. Our gestation length is only 9 months instead of the extrapolated 14–18 months. While it is argued that the human brain merely continues its development outside the womb, the period of infancy (defined by nursing) is not extended. When gestation length and weaning age are added together, humans are nourished by their mothers for less time than the other great apes and much less than might be expected (Fig. 3). Instead, humans have a unique period of dependency called childhood, in which they must continue to be fed and protected by other members of the social group for a considerable period after weaning. Individuals who stand in for parents and assist with childcare are called alloparents. Other

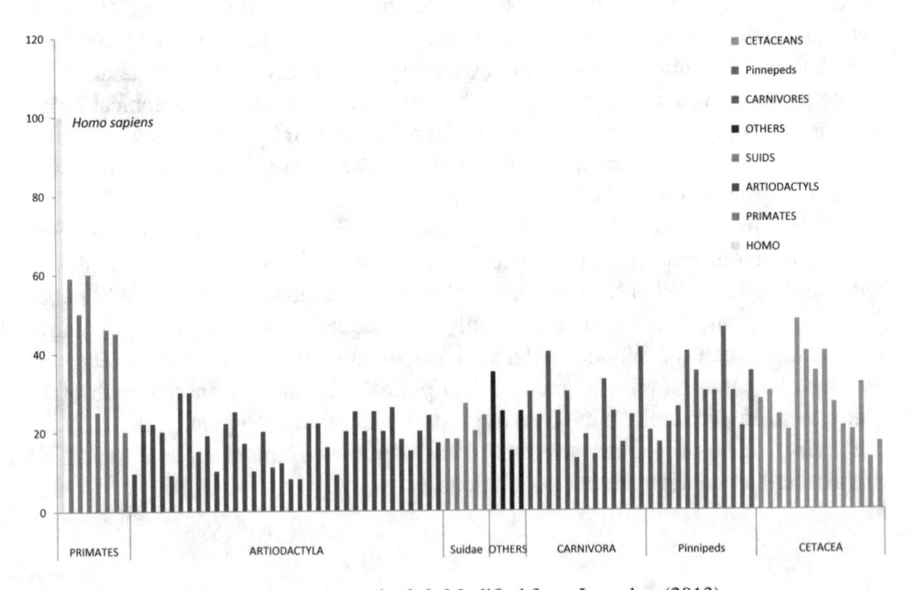

Fig. 1 Longevity. *Homo sapiens* is on the *left*. Modified from Langdon (2013)

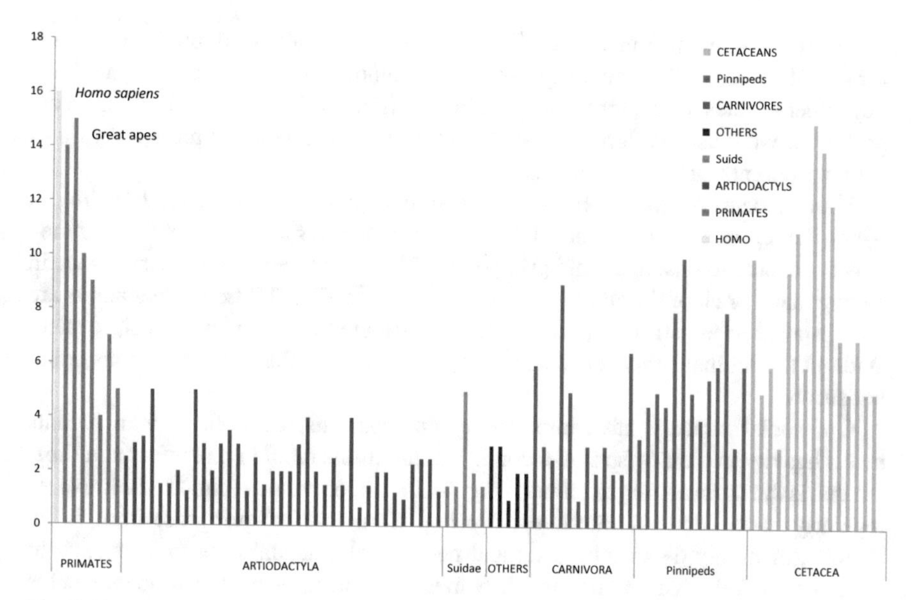

Fig. 2 Female maturation age from conception. *Homo sapiens* is on the *left*

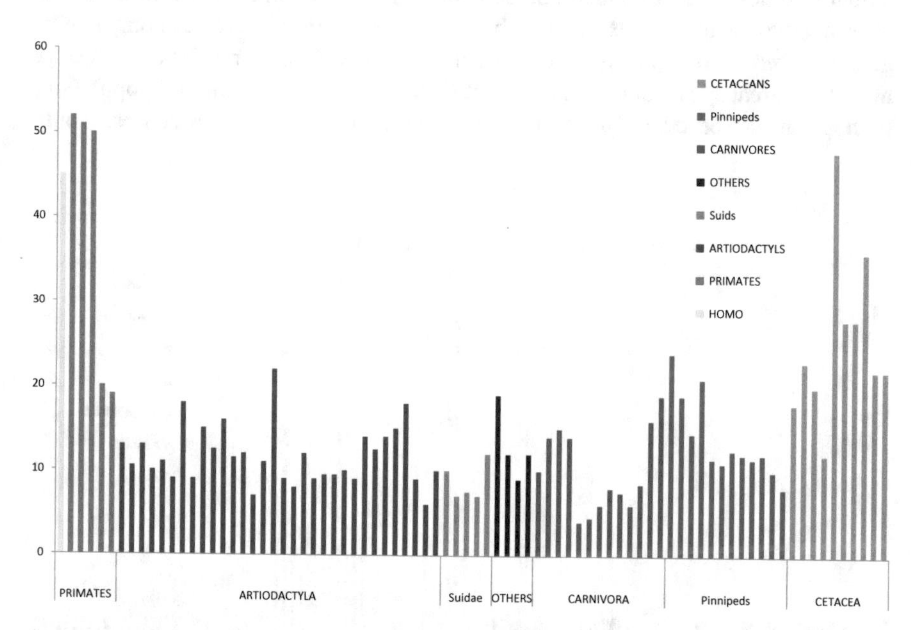

Fig. 3 Development: gestation length + weaning age. *Homo sapiens* is on the *left*. Modified from Langdon (2013)

mammals may live within a group for protection and to learn essential social skills for a while before sexual maturity, but if they cannot feed themselves, they will die. Few other species have alloparents. The implications of this strategy for humans are profound. We must develop in a social context and we depend on more individuals than our parents alone in order to thrive.

Many species of insects, birds, and mammals practice cooperative breeding in which the reproduction of some adult members is suppressed or delayed so that they may contribute to raising the offspring of a dominant breeder. Humans are unique in having many or all of the adult females capable of bearing children. Other adults are not enslaved to assist; instead the work is distributed and support is reciprocated. Without the mutual support and economic exchanges of the society, humans could not survive.

Our social strategy has reproductive repercussions, as well. Human parents invest heavily in their offspring, a condition that limits them to one infant at a time, but that is the norm among mammals (Fig. 4). In fact, only suids (pigs) and carnivores regularly have litters of multiple offspring. Carnivore offspring are born in an altricial, or highly dependent, state and need developmental time to learn how to hunt successfully. The ability to birth many offspring at a time probably helps carnivore populations respond rapidly to fluctuations in the prey abundance. While humans do not have more than one baby at a time, they can take advantage of the care shared with alloparents and have babies closer together (Fig. 5). Among hunter-gatherers, births are typically spaced about 4 years apart, similar to the reproductive rate of the great apes, but less than expected. In a sedentary agricultural population, birth spacing can be reduced to 2 years or even, under exceptional circumstances, to one.

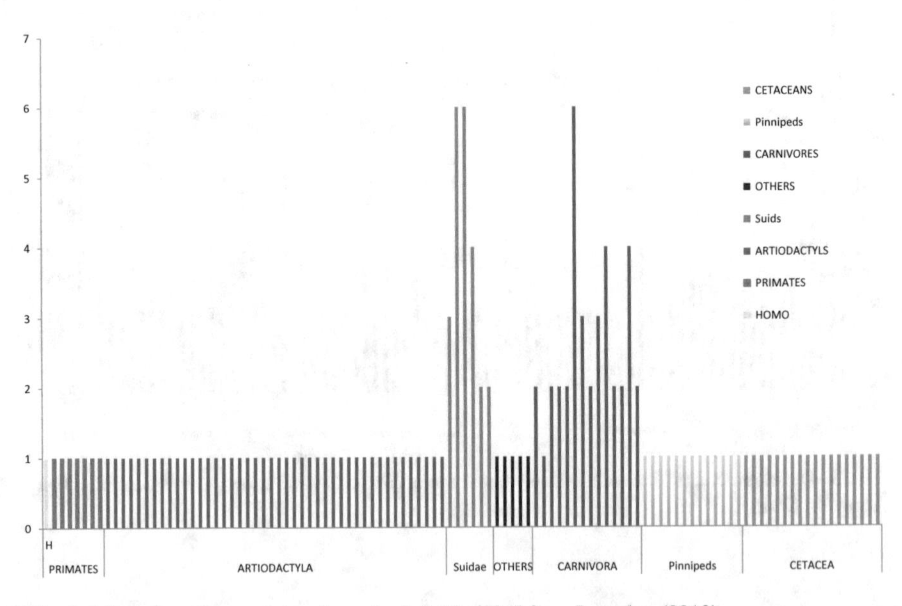

Fig. 4 Litter size. *Homo sapiens* is on the *left*. Modified from Langdon (2013)

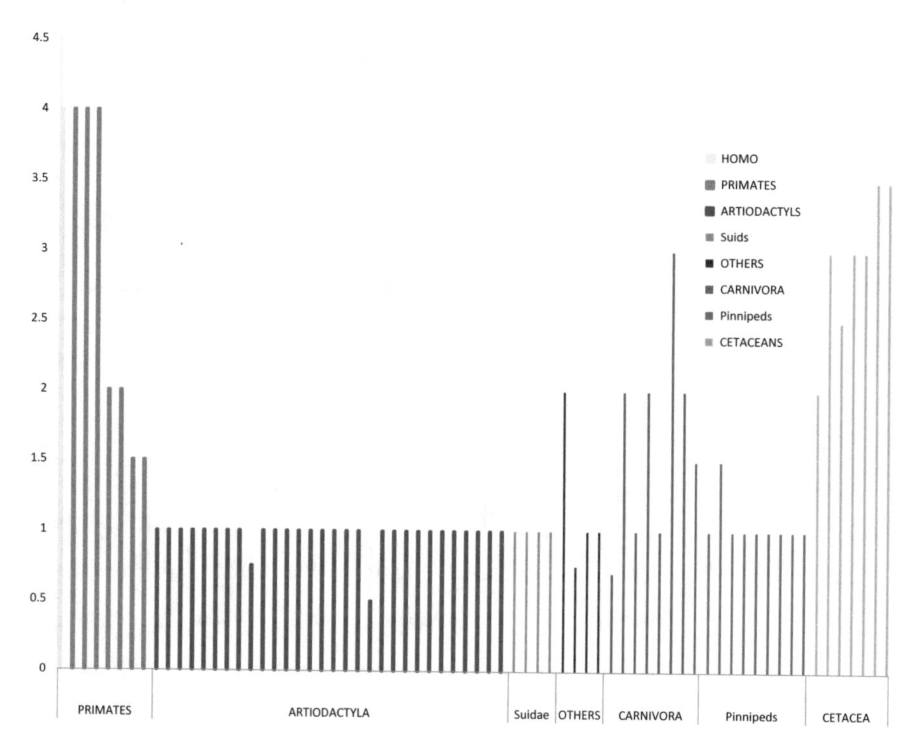

Fig. 5 Interbirth interval. *Homo sapiens* is on the *left*. Modified from Langdon (2013)

This means that, even though it takes a long time for individuals to reach maturity and begin reproducing, well-fed mothers have the potential fertility to increase the size of a population very quickly.

In summary, human life history strategies are best modeled by the great apes, but our social complexity reduces the burden on individual mothers and allows them to increase their fertility. How far back in human evolution did alloparenting begin to have such an impact on our life history? This is difficult to determine in the absence of direct evidence. The Nariokotome skeleton and possibly Neanderthals show a somewhat faster rate of development than modern humans, but it is possible that a gradual slowdown had already started by then.

Dietary Breadth

Dietary breadth can be described by the number of food groups a genus consumes. Because few species have been observed as closely as humans, dietary breadth as inferred from the literature must be underestimated for many mammals. Nevertheless, of 19 categories of food items reported for mammals, humans

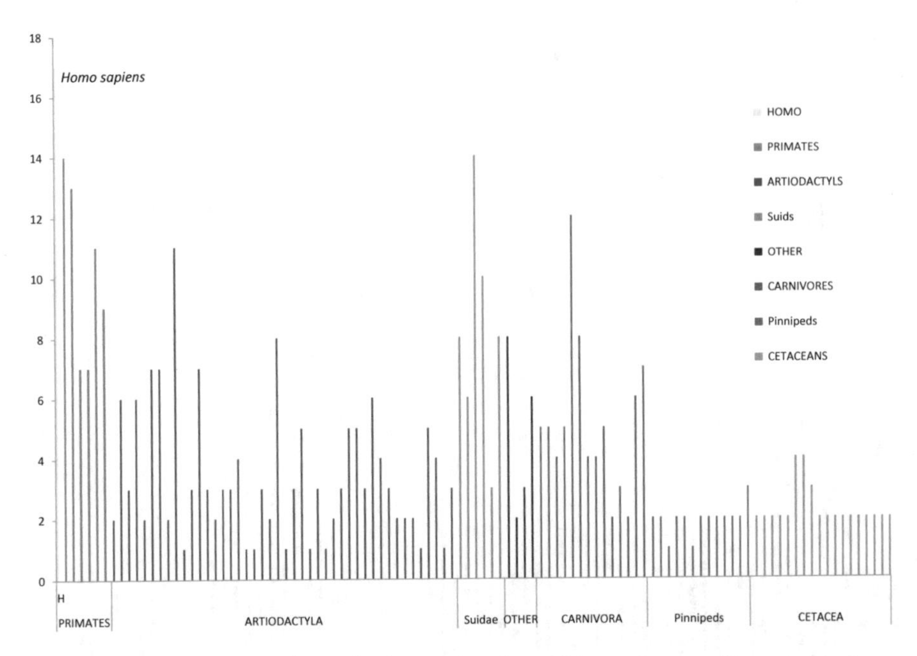

Fig. 6 Dietary breadth: Number of food categories observed for each genus. *Homo sapiens* is on the *left*. Modified from Langdon (2013)

are known to consume all of them, and 17 with some regularity (Fig. 6). Other species considered omnivores, particularly chimpanzees (*Pan*), pigs (*Sus*), gorillas (*Gorilla*), and bears (*Ursus*), have only been reported to eat from 12 to 14 of them. It is reasonable to assert that no other mammal has a broader diet than humans and few are close.

While evidence exists of certain foods being consumed by our fossil ancestor, it is not possible to rule out any category as never being consumed. All we can do is point to the earliest evidence of certain items (Table 1). If we assume our last common ancestor was chimp-like in regard to diet, then the earliest hominins were already exploiting a wide range of food sources, including fruits, leaves, insects, and other small animals. This hypothesis is consistent with interpretations of australopithecine teeth and jaws. Direct dietary evidence includes processed ungulate bones starting at 2.5 Ma from various sites in East Africa, tools possibly for digging out termite nests by 1.8 Ma, and nuts at about 1.0 Ma. Other early categories include roots and tubers (inferred for *Australopithecus* before 2.0 Ma from carbon isotopes), aquatic invertebrates (inferred from presence of shells and fish bones from the earliest Pleistocene in East Africa), marine animals (200 Ka in South Africa), fungi (12 Ka at Star Carr), and a variety of domestic crops starting about 10, 000 years ago.

Humans can be very efficient at exploiting prey items, as indicated by the dietary changes in the Late Stone Age in South Africa and Upper Paleolithic in

Table 1 Evidence for hominin dietary breadth

Food category	Earliest evidence in hominin diet
Medium/large vertebrates	Analogy with great apes and modern humans. Evidence of butchering starting in the late Pliocene (Chap. 12); possibly from 3.4 Ma at Dikika, Ethiopia
Small vertebrates	Analogy with great apes and modern humans
Eggs	Analogy with great apes and modern humans. Ostrich egg shell fragments appear in MSA deposits in South Africa
Terrestrial invertebrates	Analogy with great apes and modern humans. Tools at Swartkrans show wear patterns consistent with excavating termite colonies
Carrion	Inferred from archaeological evidence from the early Pleistocene (Chap. 12)
Fish	Early Pleistocene, Koobi Fora; inferred at Olduvai
Aquatic invertebrates	Mollusks, turtles in Middle Stone Age and after; coastal resource exploitation (Case Study 19). Exploitation of shellfish probably has a much greater antiquity
Fruit	Analogy with great apes and modern humans
Browse, shoots, fobs, bryophytes	Analogy with great apes and modern humans
Mature leaves	Analogy with great apes and modern humans
Grasses, sedges	Inferred for australopithecines and early *Homo* from stable carbon isotope ratios, Pliocene
Seeds, grains, nuts	Analogy with great apes and modern humans. Evidence for nut consumption in Early Pleistocene occurs at Gesher Benot Ya'aqov
Wood, bark, stems, pith	Observed in modern populations but not documented in the past
Roots, tubers	Tubers hypothesized for earliest hominins; inferred for early *Homo,* Early Pleistocene, from strontium isotope levels
Sap	Observed in modern populations but not documented in the past
Flowers	Observed in modern populations but not documented in the past
Mushrooms, fungus	Not observed until the Mesolithic, but probably consumed earlier
Aquatic plants	Analogy with great apes and modern humans
Dung, dirt	Observed to small degree in modern populations for possible mineral content and medicinal purposes

Europe. Such changes in the prey species exploited or in the age of the prey is best understood if the people were not exclusively dependent on the animals they were overexploiting, but able to switch readily among food items as any one resource is used up.

Habitat Breadth

Habitat breadth was measured in a similar way, by counting the number of habitat types occupied by a given genus. Modern hunter-gatherers can be found in all but two of the 15 types (Fig. 7). Oceans and polar ice caps may be occupied or exploited

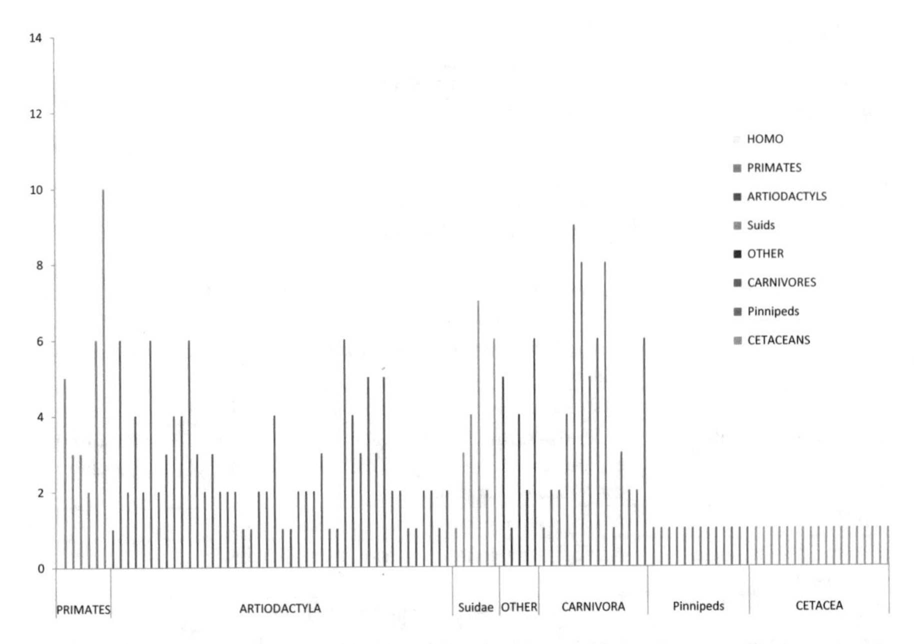

Fig. 7 Habitat breadth: Number of habitat categories observed for each genus. *Homo sapiens* is on the *left*. Modified from Langdon (2013)

for brief periods, but people cannot live in them. No other mammals are as flexible. Macaque monkeys (*Macaca*), bears, and wolves (*Canis*) are found in eight or nine different habitats. Carnivores are more likely to span several types of ecosystems, since their ability to hunt and process prey is less dependent on the environment than would be expected of herbivores.

Human ecological adaptability can be observed far back into our past (Table 2). Our ancestors occupied at least nine different habitats by the Middle Pleistocene and had spread from tropical to temperate latitudes. As has been discovered, the first members of our genus tended to prefer mosaic habitats that offered a greater range of resources. By the Middle Paleolithic or Middle Stone Age, people had moved into desert and tundra. Evidence for humans in rainforests and taiga, two of the least productive ecosystems, comes later.

Our adaptability is also reflected in geographic spread across the earth. Hunter-gatherers may be found from the far north above the Arctic Circle to the tip of Patagonia, a range of 128° of latitude (Fig. 8). Only dolphins range as far, but obviously they are restricted to a single watery habitat. Some large carnivores have spread far above and below the equator. Wolves span 114°, and cougars and the other large cats over about 105°. Again, the hunting niche is less sensitive to habitat and climate. Human hunting is probably an important adaptation that allows us to move so readily into new and different areas.

Table 2 Evidence for hominin habitat breadth

Habitat	Early evidence of occupation
Rainforest	By Late Pleistocene, e.g., Niah Cave, Borneo; Batadomba-lena, Sri Lanka
Tropical forest	By Middle Pleistocene, e.g., Bodo, Ethiopia
Deciduous forest	By Middle Pleistocene, e.g., Gesher Benot Ya'aqov, Israel; Bose Basin, China; Balkans
Taiga	Modern hunter-gatherers
Scrub forest	By late Middle Pleistocene, e.g., Duinefontein 2, South Africa
Savanna	By Early Pleistocene, e.g., Swartkrans, South Africa; West Turkana, Kenya; Yuanmou, China; Wonderwerk Cave, South Africa; Doornlaagte, South Africa.
Steppe	By Early Pleistocene, e.g., Dmanisi, Georgia
Tundra	By Middle Paleolithic, e.g., Mamontovaya Kurya, Siberia
Desert	By Middle Stone Age, e.g., Nubian Complex, northeast Africa
Mountains	By Middle Pleistocene; Caucasus, e.g., Achalkalaki
Swamp, wetlands	By Middle Pleistocene, e.g., Buia, Ethiopia; Olorgesailie, Tanzania
River/lakeside	By Early Pleistocene, e.g., Olduvai Bed II, Tanzania; 'Ubeidiya, Israel; Nihewan Basin, China
Coast	By Middle Pleistocene, e.g., Boxgrove, England; Terra Amata, France
Ice cap	Proposed in late Pleistocene
Oceans	Not inhabited by hominins

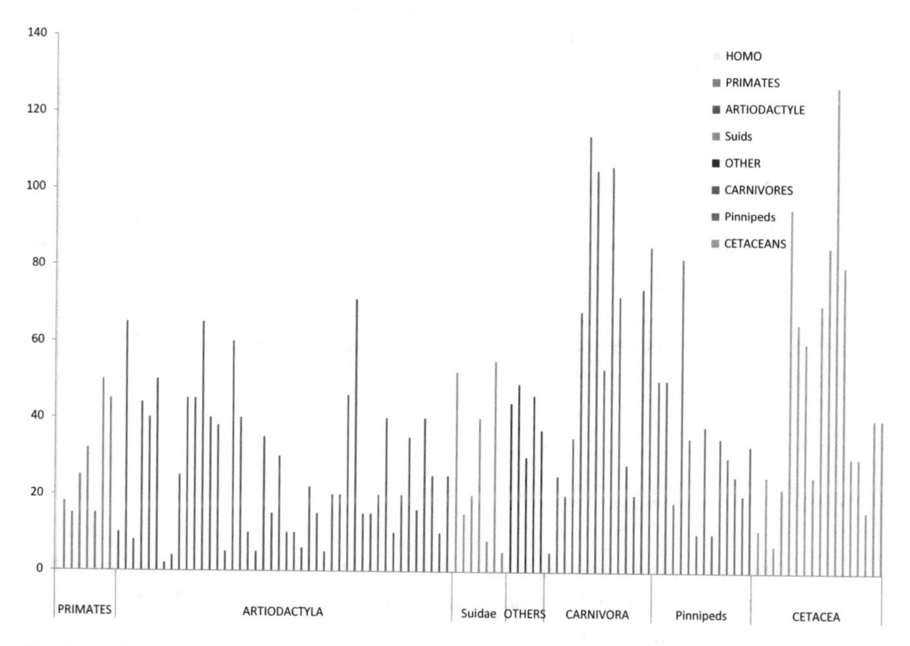

Fig. 8 Latitudinal range. *Homo sapiens* is on the *left*. Modified from Langdon (2013). Source: Langdon, JH 2013. Human ecological breadth: Why neither savanna hypotheses nor aquatic hypotheses hold water. Human Evolution 28(3–4):171–200

Table 3 Geographic and latitudinal range expansion of genus *Homo*

Date	Site	Range (°)
Africa, Late Pliocene before 1.8 Ma	Ethiopia: Gona, Bouri (11°N)	21
	Malawi: Uraha (10°S)	
Africa, Early Pleistocene ~1.8 Ma	Algeria: Ain Hanech, El-Kherba, Algeria (37°N)	63
	South Africa: Swartkrans, Sterkfontein (26°S)	
Asia, Early Pleistocene to 1.2 Ma	Georgia: Dmanisi (42°N)	68
	Pakistan: Riwat (37°N)	
	China: Majuangou (40°N)	
	Java: Modjokerto (7°S)	
Middle Pleistocene to 0.8 Ma (water crossings)	South Africa: Pneil 6 (28°S)	80
	China: Gongwangling (40°N)	
	Spain: Orce, Fuente Nueva, Atapuerca (42°N)	
	Italy: Isernia, Monte Poggiolo, Ceprano (44°N)	
	Germany: Dorn Durkheim (50°N)	
	England: Happisburgh (52°N)	
Middle Pleistocene to 0.4 Ma	South Africa: Saldanha (33°S)	85
	China: Jinniushan (41°N)	
	Korea (38°N)	
	Tajikistan: Khonalo, Knonako, Kuhi-Pioz (38°N)	
Middle Paleolithic to 40 Ka	South Africa: Klasies River Mouth (34°S)	102
	Russia: Garchi, Diring-Yuriakh, Olalinka, (61°N)	
	Yenisei Bay (72°N)	
Early modern cultures	Australia: Keilor (38°S)	110
	Sapochnaya Karga (72°N)	

Before 2.0 Ma, *Homo* was found only along the African Rift Valley, but within 200,000 years, humans had traveled with remarkable speed from South Africa to Northwest Africa the Caucasian Mountains (Table 3). Soon after, they showed up in Northwest China in the Nihewan Basin and in Java. Although glaciations during the Pleistocene made much of Eurasia uninhabitable for most of that period, there is evidence that humans extended their range northward when possible, including in England 800 Ka and Siberia by 45 Ka.

Ecological Strategy and Sustainability

What are the appropriate animal models for human nature? As we have seen, our life history strategy is a recognizable extension from that of the great apes, as is our wide-ranging omnivory. In our ability to thrive in a variety of habitats, we resemble the social carnivores, particularly wolves. This combination of a broad diet and broad habitat exploitation is a powerful mix. Clearly, humans show an extraordinary adaptability to find resources and tolerate the elements in nearly any habitat. The keys to this success are equally apparent—an intelligence that can appraise new challenges and solve problems; material culture that is cumulative and inventive; and a social support system that shares economic resources and cooperates to achieve ends not possible for a single individual. These skills allow humans to overcome ecological barriers and expand our population geographically even as we increase its density.

There is also evidence of our own impact on the environment. While today society worries about overfishing, deforestation, and loss of biodiversity, such outcomes were already apparent in by the Upper Paleolithic when humans had to subsist on smaller prey that was more difficult to catch (Case Study 18). People have controlled fire perhaps for a million years. Hunter-gatherers discovered that it burns away brush and slows the renewal of forests, opens up grassland, and attracts herd of grazing game animals. About 45,000 years ago, fire was being used to alter habitats in Australia where it contributed to the extinction of most of the large species of mammals. Fire, reshaping of habitats, pressure on prey populations, and extinctions—these are also indicators of the human ability to overcome ecological barriers to population growth. However, as the population has grown to modern proportions, the pressures it has placed on the environment have become unsustainable.

Nothing in this portrait of humanity is unique by itself. Even material culture exists at rudimentary levels in the apes and a variety of other animals. We are not unusual in our desire to reproduce and expand our population, simply more successful at it. When did humans become unsustainable? Was it early in the primate lineage when complex societies became the norm? Was it when our brains began their long path of expansion 2 Ma ago? Or was it when the human population density reached a critical point that accelerated technological progress perhaps 60,000 years ago? Has there ever been a time in the past, an ecological Rubicon, when evolution could have stopped and left us forever as innocent as every other species, but equally at the mercy of the environment?

Questions for Discussion

Q1: What is meant by "human nature"?

Q2: What is the nature of humans? Suppose you were an extraterrestrial observer. How would you describe human behavior in ways that invite comparison with other species, such as chimpanzees or horses?

Q3: Imagine placing a social group of elephants or tigers or zebras in an unfamiliar habitat. Would you expect to thrive? Why or why not?

Q4: Populations of other species are kept in check by food supply or other limiting resources. Is that also true for humans?

Q5: Do all species have the tendency to be unsustainable if they could? Why is this an issue we only discuss in terms of humans?

Additional Reading

Fowler CW, Hobbs L (2003) Is humanity sustainable? Proc R Soc Lond B 270:257–2583
Hrdy SB (2011) Mothers and others. Belknap, Cambridge
Langdon JH (2013) Human ecological breadth: why neither savanna hypotheses nor aquatic hypotheses hold water. Hum Evol 28(3–4):171–200
Potts R (1998) Variability selection in hominid evolution. Evol Anthropol 7:81–96
Wells JCK, Stock JT (2007) The biology of the colonizing ape. Yrbk Phys Anthropol 50:191–222

Case Study 24. The Unknowable Biped: Questions We Cannot Answer

Abstract The fossil and archaeological record has revealed much yet remains frustratingly incomplete. Moreover, there are important questions it can never answer. Anthropologists may learn what species existed where and when; they can formulate hypotheses about what earlier hominins ate, how they grew up, and how their bodies changed. Science has revealed something of their environmental conditions and the other species with which they interacted. The fossils reveal how our bodies have evolved, but they cannot answer the question of why these changes took place.

Humans are unique, but how did we get that way? Why are we bipedal? Why did our canine teeth become smaller? Why do humans cooperate socially? Why do we have language while no other species does? Why indeed do we have an intellect capable of asking such questions? These are questions about human nature and our identity as a species. Despite continuing study and speculation and many hypotheses, science is no closer to certainty in answering these "why" questions than it was 200 years ago, and there are good reasons it never will be.

The Enigma of Bipedalism

Bipedalism is one of the hallmarks of humans among living species and a defining feature of hominins in the fossil record. It can be recognized in the skeletal remains. Bipedalism helps to define the hominins; thus by definition, it was one of the first human traits to evolve. Descriptions of human uniqueness commonly begin with upright posture.

Why are humans bipedal? Few other animals are, and the examples available (e.g., kangaroos, birds, dinosaurs) are so different anatomically, biomechanically, and behaviorally that they offer few insights into the hominin transformation. Many hypotheses have been proposed, and they represent several different lines of inquiry. All assume that bipedalism is adaptive and has been the outcome of natural selection. To assume otherwise leaves nothing on which to build a story. However, the barriers of transforming from a quadrupedal design to upright posture seem

J.H. Langdon, *The Science of Human Evolution*,
DOI 10.1007/978-3-319-41585-7_24

insurmountable. Humans are slower and more prone to falling and injuring themselves than are four-legged animals, and any intermediate state that has been imagined appears to have put our ancestors at an even greater disadvantage. Either the selective forces involved must have been very strong or—more likely—the circumstances were different than generally imagined.

Other Uses for Hands

The most common approach to this problem has been to ask, "What do humans do with their freed hands?" Perhaps some of these activities led of habitual bipedalism, but choosing among the possible answers to this question becomes a debate about human nature and is interwoven with social values.

Using tools. Charles Darwin argued that we became bipedal so that our hands were free to make and use tools. Increased use of tools participated in a positive feedback to make us increasingly committed to bipedalism. At the same time feedback between the brain and use of tools contributed to greater intelligence and greater dependence on technology. Darwin wrote in a society and an era when technological progress seemed capable of solving all human problems. In that context, technology was a large part of his understanding of what it meant to be human.

Using weapons. Raymond Dart's vision of humanity was much more bleak (Case Study 8). His imagined Killer Ape stood upright to wield tools as offensive weapons, employing its hands for hunting food and confronting one another.

Defense against predators. The savanna can be a dangerous place. Humans without culture are defenseless against large cats and packs of hunting dogs and hyenas. Adriaan Kortlandt proposed that by waving bundles of thorn branches and throwing rocks, hominins may have been able to keep predators at bay while group members butchered a carcass. Louis Leakey successfully tested this concept, but against East African lions who already associated humans with spears and guns.

Carrying food. One of the striking differences between humans and most other mammals is our tendency to bring food home and to share it. It has been suggested that a carrying device may have been the first important tool, but hands would have served before then if they were not needed for locomotion. This argument was proposed by Gordon Hewes in the 1950s as anthropologists were becoming more interested in primate and hunter-gatherer models for our ancestors and assumed that families, division of labor, and "home" characterized our ancestors just as they do modern humans.

Carrying babies. When hominins lost their body hair, our infants lost their ability to cling to their mothers. Human infants are especially helpless, and it would have become increasingly necessary for their mothers to have arms free to hold and carry them during gathering expeditions. Jane Lancaster and others presented this argument in the 1970s amid challenges to the male-oriented Hunting Hypothesis. As the feminist consciousness transformed the perceived role of women from dependent wives to active providers, prehistoric women made a similar transition.

Provisioning females. C. Owen Lovejoy proposed that males who carried food home to females would have improved the health and fertility of their mates and therefore increased their own reproductive success. Lovejoy published his argument under the title "The Origin of Man" in 1980 and combined that argument with explanations of pair-bonding, hidden estrus, and other traits. At that moment in American history, closely following the peak of feminist movement, it is surprising that he depicted gender roles in such an asymmetric fashion, but it was also a time of growing concern about the integrity of the family and of conservative resistance to the liberalism of the preceding two decades.

Nonhuman Bipedalism

Another approach to the problem of bipedalism has been to ask, "When are other primates bipedal?" All primates are capable of standing and walking on two feet for brief periods. If those behaviors became more important among our ancestors, that might explain bipedalism.

Display and intimidation. Gorillas famously stand and beat their chests to intimidate rival males. Chimpanzees may scream and jump around, waving branches to amplify the commotion. Display and intimidation are safer alternatives to physical conflict for exercising power and controlling reproductive opportunities, and upright posture increases apparent body size to impress rivals.

Foraging. Monkeys, apes, and even nonprimate mammals have been observed standing erect to pick fruit from low branches and bushes. This behavior accounts for the largest number of observations of bipedal monkeys and chimpanzees. It also coincides with more foraging from ground level and might have signaled an ecological shift in our ancestors from greater arboreality to greater terrestriality.

Observation. Savanna grass can be tall—too tall for a smallish quadruped to see over. Standing upright might provide early warning against predators and simply help to keep tabs on the rest of the social group.

Wading. Chimpanzees, gorillas, and orangutans have all been observed to wade bipedally in shallow water. Perhaps this was to hold their heads out of deeper water or because they didn't want to get any wetter than necessary. Primates appear to have the same ambivalence about getting wet that humans have. At different times, chimpanzees have been observed bathing to cool off, but also cowering miserably under trees during a rainstorm.

Locomotor Models for Our Ancestors

A third approach asks, "Which primates could most easily make the transition to obligate bipedalism?" All primates sit upright and occasionally take a few steps on two feet, but which species better model our ancestors? Answers to this question might help us understand how the transition occurred, with the implicit understanding that a transition from trees to the ground was the driving selective force behind it.

The hylobatian hypothesis. Gibbons swing gracefully by their arms when in trees, but walk bipedally on the ground. Their upper limbs are too long to be used effectively there and are not built for weight-bearing. Could such behavior have described our ancestors? Unfortunately, their lower limbs show no compromises for walking and it remains awkward and inefficient. Gibbons much prefer to remain high in the trees.

Brachiating hypothesis. Brachiation—arm-swinging through the trees—puts the body in an upright position and leads to a specialization of the limbs. The upper limb is used for suspension and the lower for weight-bearing. Even though the gibbon condition was dismissed as too specialized, for a long time living apes were all classified as brachiators (even gorillas) and it was assumed the condition described our common ancestor too.

Climbers. A more sophisticated understanding of living apes recognized that their distinctive body posture, limb proportions, and trunk design were better explained by climbing behaviors. Unlike smaller monkeys, apes tend to hoist themselves up with their hands. Numerous climbing adaptations can been seen in the human skeleton and even more existed in early hominins.

Efficiency Experts

A key study by Peter Rodman and Henry McHenry found that humans on a treadmill are at least as efficient as chimpanzees, whether bipedal or quadrupedal; previous comparison with nonprimates was misleading. Chimpanzees have adaptations of the hand, wrist, and elbow for a form of quadrupedalism called knuckle-walking. Strengthening of those joints increases their ability to tolerate greater forces, but chimpanzees are still not very efficient walkers and runners. While studies of horses and similar quadrupeds made humans look so slow and ineffective that the evolutionary switch to bipedalism appeared highly improbable, Rodman's study tells us those are false comparisons. If our last common ancestor with chimpanzees was generally chimp-like, as is commonly assumed, perhaps without the knuckle-walking modifications, the barriers to hominin bipedalism would not have been so great. Natural selection merely favors improvements over the present condition. We can now focus on the question of energetic efficiency: "When is bipedal efficiency important?"

Sustained walking. Humans have great endurance, enabled by adaptations throughout the body (Case Study 14). Today's hunter-gatherer women may walk many dozens of kilometers each week carrying food, firewood, and children. Hunters travel to find game and may track animals for days. Entire bands relocate on a regular basis when local resources become exhausted and commonly trek to larger social gatherings. In parts of the world, including East Africa herds of game animals migrate thousands of kilometers seasonally, and their predators often follow them. Such behavior patterns were probably important in the past and led to population of the surface of the globe. It must have placed a selective pressure on hominins to be able to fuel their exertions and keep up with their social groups.

Running. Humans are also capable of sustained running when speed is more important. This may have critical to facilitate hunting.

Thermoregulation. Peter Wheeler calculated that upright body posture exposes less surface area to the noon sun and more to cooling breezes. The tropical heat drives most mammals to shade and inactivity in the middle of the day, but our superior ability to dissipate excess heat may have opened new ecological niches.

Wading. If our ancestors spent much time foraging in water of a certain depth, upright posture would have increased efficiency. Some researchers have argued that shellfish in the East African lakes and later along the coast may have been an important resource for early hominins.

No Answers

The ideas listed above summarize most of the hypotheses taken seriously by anthropologists at one time or another, but many have been left out. Algis Kuliukas listed 42 specific, though often only subtly different, explanations for bipedalism. Clearly there is no dearth of ideas for why we walk on two feet. We are unique. We must be the products either of a unique selective pressure or a unique environment or both. Why, then can't we answer such an important question? There are several reasons.

First, these hypotheses are untestable. Observational and experimental data can and have been collected. We know how frequently and under what circumstances chimpanzees, baboons, and other primates stand and walk bipedally. We can measure the relative energetic efficiency of different activities and modes of travel in different species and can even model alternative anatomies. But our question asks about the relative importance of activities in the past, not the present; on this we can only speculate. The evolution of bipedal hominins occurred only once and will not happen again. Scientists cannot create experimental situations where some variables are controlled and others altered; therefore, the causes of past changes cannot be tested. This is a limitation that applies to all of evolutionary history. It prevents us from resolving the question of whether our evolutionary path was determined or alternatives might have been possible, and it prevents us from making significant predictions about the evolutionary future.

Second, the origin of bipedalism was a process, not an event. Evolutionary change in a population with a long generation time takes place over thousands of years, at least. It results from the differential success and survival or premature death of many individuals and lineages. Even with a time machine that could place us in any date and place, what would we look for to answer such questions? At best we would have one or a series of snapshots of how hominins moved about and in what environments. Those answers we might get from the fossil record. Tracking selection would require an immense set of measurements of the lifetime success of many individuals and their offspring.

Third, we now know that bipedalism did not come about in a single event. Australopithecines were bipedal with a pattern that persisted for 2 Ma or more. Why they became bipedal is the first problem. But *Homo* has a different body design, apparently more suited to habitual terrestrialism. At least a second event took place that needs a second explanation. The circumstances and explanations for these two (or more) stages of locomotor change likely were very different. Early australopithecines and their ancestors lived in more wooded landscapes than *Homo*, probably used fewer tools, and were less concerned with animal protein. Humans occupied a greater variety of landscapes, scavenged and/or hunted, and became dependent on technology and culture. The selective pressures and the interactions between locomotion and ecological niche were certainly different. No single model will explain bipedalism, and studies of modern human locomotion have limited relevance to australopithecines.

Finally, there is no reason to believe a single factor was operating at any time during either of these transitions. We cannot begin to understand the process without knowing what our prebipedal ancestor was like. It has been assumed for most of the last century that ancestor was an arboreal climber similar to a chimpanzee in many ways. Owen Lovejoy's reconstruction of *Ardipithecus* as an arboreal quadruped that was bipedal on the ground has directly challenged that assumption. If he is correct, anthropologists face an even greater challenge in explaining the change in posture. However, Lovejoy's argument fails to account for the many vestiges of climbing adaptations in fossil and modern anatomy, including the shape of our ribcage, orientation of the hominin shoulder, and ape-like proportions of limb lengths in australopithecines. These features favor a climbing model for our ancestor and suggest that *Ardipithecus'* proposed position as a near cousin of our ancestor is incorrect. Such starting conditions of our lineage, what Stephen Jay Gould called "historical contingencies," would have constrained what was possible and what was a likely evolutionary response to new challenges.

Accepting chimpanzees as the best available model for our climbing ancestor, we note that they often stand and briefly walk on two limbs. How frequently did our ancestor assume an upright posture? At what point would they be considered bipedal? In those initial stages, perhaps efficiency was the most important selective force, but that animal was also climbing and foraging. Over time diet, land use, and social behavior were evolving. All of them interacted with posture and locomotion. How can one expect to isolate one or even a few factors and conclude that they explain why bipedalism evolved?

Anthropologists assume bipedalism is an adaptation. What if that explanation is not entirely correct? It is difficult to imagine that such a thorough reorganization of the body might have come about through genetic drift or some other random process; yet there may have been a time during which our climbing ancestor found advantages to spending more time on the ground but had no adaptations for it. No living apes can compare in locomotor efficiency to ground-dwelling species. Orangutans, like gibbons, spend most of their time in trees and make do on horizontal surfaces with what they have. Gorillas and chimpanzees knuckle-walk, with reinforcements of the elbow, wrist, and hands to bear weight. Like australopithecines, chimpanzees show a compromise between climbing and terrestrial movement, but chimpanzees appear less committed anatomically to the ground. The last common ancestor of humans and chimpanzees may well have gone through a phase with few adaptations to the ground, a time when bipedalism was no better or worse a choice. Did culture or social behavior play a role in pushing us one direction or another? Or was the direction our line took merely fortuitous? Alternatively, was there something in our ancestor's anatomy that made bipedalism an easier path? From that point on, our posture interacted with all the ecological, social, and cultural activities that defined the hominin niche. Once the ancestor was even slightly inclined to bipedalism, selective pressure would have been continuous to improve efficiency and their ability to exploit opportunities that the environment presented. The rest, one might say, was evolutionary history.

Questions for Discussion

Q1: How can one decide that one explanation of bipedalism is better than another?

Q2: If the question cannot be answered, what value is there in debating possible answers?

Q3: Often we can make more progress by asking better questions. What other questions can one ask to help us understand why we are bipedal?

Q4: How important is it that a model be able to make predictions, rather than explain observations after the fact?

Q5: Is it justifiable to simplify and downplay scientific controversy in order to communicate more effectively with the public, even if it means presenting a false appearance of certainty? Would you answer the same way if the question were about political decisions rather than about science?

Q6: What important human traits, in addition to bipedalism, have not been explained? Are they explainable?

Additional Reading

Darwin C (1871) The descent of man and selection in relation to sex. Reprinted Modern Library, New York

Fleagle JG et al (1981) Climbing: a biomechanical link with brachiation and with bipedalism. Symp Zool Soc Lond 48:359–375

Kuliukas AV (2011) A wading component in the origin of hominin bipedalism. In: Vaneechoutte M et al (eds) Was man more aquatic in the past? Fifty years after Alister Hardy: Waterside hypotheses of human evolution. Bentham eBooks, Oak Park, pp 36–66

Lovejoy CO (1981) The origin of man. Science 211:341–350

Lovejoy CO (2009) The great divide: *Ardipithecus ramidus* reveals the postcrania of our last common ancestors with African Apes. Science 326(73):100–106

Morris D (1967) The naked ape. McGraw-Hill, New York

Rodman PS, McHenry HM (1980) Bioenergetics and the origin of hominid bipedalism. Am J Phys Anthropol 52:103–106

Case Study 25. Parallel Paradigms: Umbrella Hypotheses and Aquatic Apes

Abstract An umbrella hypothesis is an evolutionary scenario built around a premise that offers answers for a wide range of adaptive problems. Umbrella hypotheses have a deceptive appearance of parsimony because the explanations they offer for independent characters are hypotheses that are no better tested or supported with evidence than before. Instead, another layer of untestable hypothesis has been added. The example examined in this chapter is the Aquatic Ape Hypothesis, which proposes that important human characteristics were originally adaptations to an ecological niche in or near the water. Unlike many conventional umbrella hypotheses, this one emerged within a separate paradigm that has not yet been reconciled to mainstream paleoanthropology.

Umbrella Scenarios

It is very easy to spin elaborate scenarios of human evolution. The case studies in this collection refer to several such examples: the Killer Ape, the Hunting Hypothesis, the Savanna Hypothesis, the Scavenging Hypothesis, and various climate models. Still others have circulated in the discipline, including sexuality, cooking, self-domestication, and an aquatic phase as proposed prime movers of evolution. Each of these starts with a premise from which speculation flows freely and generally offers explanations for the most important traits that define our species. For example, the Savanna Hypothesis first set out by Darwin explained bipedalism, leading into a feedback loop involving tool use, intelligence, and canine reduction. The Hunting Hypothesis went beyond this to account for a carnivorous diet, social organization, sexual division of labor, tribalism, and language, as well. Such hypotheses have the appearance of parsimony because they appear to explain so much: If one accepts the underlying premise, everything else follows in a tidy narrative.

The reality is that each trait "explained" is just another hypothesis to be tested. For example, hunter-gatherers do divide labor by gender, although not always in the same way. There are many possible ways to relate economic division to a hunting and gathering life style. Perhaps a sexual division of labor appeared because

J.H. Langdon, *The Science of Human Evolution*,
DOI 10.1007/978-3-319-41585-7_25

women did not have the strength or stamina to hunt. Perhaps their role as forager and child-bearer was more important and interfered with developing hunting skills. Perhaps by not participating in the hunt, their absence facilitated male bonding and reduced sexual competition. The Hunting Hypothesis does not choose among these possibilities or give us a definitive answer for this or other traits it purports to explain. Rather than parsimoniously reducing the number of hypotheses with which we are working, it has unparsimoniously added an assumption about hunting. The new assumption alters the landscape on which we still debate adaptive scenarios for individual traits, but does not itself test them or provide answers for our questions.

In a 1997 paper, Langdon coined the term "umbrella hypotheses" to refer to these overarching models that appear to explain much while only adding speculations. The popular literature abounds with umbrellas, and few anthropologists can avoid becoming attracted to one or more of them. They provide what the discipline is seeking: a narrative of human origins. However, those stories are inevitably embedded with our own preconceptions of human nature. Victorian England had the self-assurance of an empire at its peak and Darwinism presented it with a depiction of humans utilizing culture to evolve toward perfection. The Lost Generation despaired of humanity and Dart and Ardrey gave that pessimism an anthropological voice with the Killer Ape. More recent social trends, including environmentalism, the sexual revolution, feminism, its conservative backlash, and health foods diets are all reflected in a colorful array of paleoanthropological umbrellas. Umbrella scenarios may be the best medium for communicating the excitement of the field to a popular audience. However, anthropologists do a disservice to the readers and the discipline if they fail to acknowledge the fundamental weakness and limitations of such story-telling.

Many umbrella hypotheses operate within the disciplinary paradigm, drawing upon the same body of evidence to support or test propositions or to construct new hypotheses. A few lie well outside the realm of science, invoking extraterrestrial aliens or paranormal phenomena, or simply operating with their own rules of evidence. Some however, occupy an intermediate ground, attempting to pursue science within a different paradigm. The Aquatic Ape Hypothesis is one such example.

The Aquatic Ape

In 1960, marine biologist Sir Alister Hardy published a speculative paper titled "Was Man more Aquatic in the Past?" This was the first English version in print of what has become the Aquatic Ape Hypothesis, although similar ideas had been proposed earlier by the German pathologist Max Westenhöfer. Hardy's brief account noted a number of similarities between humans and the aquatic mammals he studied: the ability to hold one's breath, the attraction beaches have for people, the loss or reduction of most of body hair, the orientation of vestigial body hairs, a streamlined body shape, and deposits of subcutaneous fat. He suggested this may all be explained if our ancestors passed through an aquatic phase that began with wading and foraging for shellfish and invertebrates along the ocean shore. Upright posture for wading led to full bipedalism, which also presented a streamlined "boat-like" profile for effective swimming. Our opposable grip would have been useful for foraging, as well as tool-making, but these later were effective for catching fish by hand.

This intriguing concept was expanded upon by the late writer Elaine Morgan, who published her account in *The Descent of Woman*. Morgan's version included a number of additional anatomical features that might be explained in the context of an aquatic phase, including tears, our protruding nose and the form of the upper respiratory system, changes in skin glands, vaginal depth, and breasts. Morgan's first book was largely ignored by the academic community and the aquatic scenario dismissed with little discussion in print. Disappointed, but not discouraged, Morgan continued to write articles and more books on the subject, including *The Aquatic Ape Hypothesis* (1997). Since the 1960s the hypothesis has gained a small body of supporters who continue to publish articles and books, but few of these researchers have risen from within anthropology.

Why was the Aquatic Ape Hypothesis dismissed by the great majority of paleoanthropologists? The silence in print falsely suggests a lack of awareness of the premise. Morgan and her supporters have accused the anthropological community of being closed-minded to challenges to the orthodox models; yet the field thrives on debate. The male-dominated field was accused of being sexist, which it has been. Although Morgan's first book assumed a combative feminist stance, that does not explain why, after 30 years, several other books and two academic conferences the Aquatic Ape model still had not received objective examination. It is true that the advocates of the hypothesis were outsiders to the paleoanthropological community, but the discipline has eagerly embraced perspectives from other scientific disciplines on many occasions.

The problem has been that from the start the standard Savanna Hypothesis and the Aquatic Ape Hypothesis were operating as different paradigms. They began with different assumptions and asked different questions. Interestingly, their arguments often converge on opposite interpretations of the same evidence (Table 1). Philosophers would describe these as theory-laden observations, in which researchers selectively focus on observations that support their expectations.

Paleoanthropologists have focused largely on the fossil record and the interpretation of the skeletal anatomy it revealed and they looked to living primates as models for ancestral anatomy and behavior. They examined the australopithecine fossils and debated what degree of arboreality they represented (Case Study 9). Anthropologists generally pay less attention to soft tissues not represented in the fossils but try to reconstruct gait patterns and mechanics. The endurance of humans in walking and running is interpreted as the result of adaptations for efficient terrestrial locomotion. Anthropologists have been quick to observe that most people either cannot swim or have to be taught, and that humans in water are always in danger of drowning.

Table 1 Theory-laden observations of paleoanthropology and the Aquatic Ape Hypothesis

Anthropological claim	Observations/evidence	AAT claim
Humans have great walking, running endurance	Modern human physiology	Humans are slow and vulnerable on the ground
Humans are poor swimmers, prone to drowning	Modern human physiology	Humans have excellent swimming, diving skills
Ancestors transitioning from arboreal to terrestrial habitat	Australopithecine fossils	Ancestors were adapted for water
Earlier hominins adapted for walking, running	Comparative musculoskeletal anatomy	
	Comparative soft tissue anatomy and physiology	Ancestors were adapted for water
Terrestrial fauna indicate hominins preferred a mosaic habitat containing grasslands	Paleoenvironmental reconstruction including fauna	Aquatic fauna indicate hominins preferred a waterside habitat
Continuous terrestrial record of hominins	Continental distribution of hominin fossil and archaeological sites	Early hominins inhabited wetlands and coastal sites now under water
Omnivorous diet with important component of meat	Archaeological evidence of hunting, butchering; modern diets	Omnivorous diet with important component of aquatic foods and occasional meat
Omnivorous diet with important component of animal foods	Modern diets; nutritional needs	Omnivorous diet with important component of aquatic foods

Supporters of the aquatic model were not engaged in paleontological fieldwork. They built arguments based on contemporary human and comparative anatomy of aquatic or semiaquatic mammals. Much of their research has been engaged with human physiology, respiratory limits, and diving abilities. Morgan argued that the transition to bipedalism would have been impossible on land because early hominins would have been too slow and vulnerable to survive. On the other hand, her followers cite humans' natural swimming aptitude. Among Hardy's evidence for an aquatic ancestry was "the exceptional ability of Man to swim, to swim like a frog, and his great endurance at it." Much is made of the skills of trained pearl divers and others with long experience near the ocean.

In the 1960s and succeeding decades, abundant fossils were recovered from the badlands of the East African Rift Valley and the cave breccia of the South African Savanna. Initially the environment in which paleontologists operated appeared to confirm an ancient grassland setting. Hominin fossils were accompanied by those of bovids and other animals documenting changing degrees of woodland and grassland through the Plio-Pleistocene. At least in the Middle Pleistocene and later, there is evidence supporting *Homo* as a big game hunter; anthropologists studied the accompanying fauna to extrapolate that niche back into the Pliocene.

Morgan and others focused on the negative evidence—the absence of fossils between the Middle Miocene and the earliest Pleistocene posed no constraints on habitat during that time period. The absence of fossils might be explained by the fact that current high sea levels are hiding the coastal habitats of early hominins.

Because most fossils are buried through the action of water, the great majority are also accompanied by bones of fish, turtles, or crocodiles, indicating the presence of a body of water. Those same waters potentially offered abundant aquatic animal and plant foods rich in the long-chain fatty acids needed by large brains. Some members of the Aquatic Ape community accepted australopithecines as living after the aquatic phase, while others interpreted them as active swimmers.

In this way, the two communities have constructed parallel paradigms in which observations are inherently consistent with and thus appear to confirm their initial premises. It should be no mystery that dialog between the two sides has been frustrating.

Waterside Hypotheses

The original Aquatic Ape Hypothesis has accumulated a diverse array of interpretations. Algis Kuliukas and Morgan summarized six competing versions in 2011, while recognizing further variants and disagreements within them, and favors the term Waterside Hypotheses as more inclusive. However, "they all share the underlying belief that aquatic scenarios are largely responsible for explaining why human beings are so remarkably different from our closest cousins, the chimpanzee." Nonetheless, they do not necessarily agree on which traits are best explained in this way. Morgan, when asked which trait lay at the core of her model, named bipedalism; yet in a hypothesis that started from analogies with marine mammals, that is one trait that has no parallels.

According to different authors, the time of the waterside phase may be in the Middle Miocene (10–15 My) or in the Middle Pleistocene (after 2.0 My). While Hardy envisioned our ancestors swimming in the Indian Ocean, Kuliukas favors a history of wading in rivers and shores of East Africa. The lakes of the Rift Valley have been proposed as potential location where hominins may have learned to forage for shellfish by wading bipedally. The lack of agreement is an indication of a paradigm that is maturing and subject to normal science. It is also what one might expect of an umbrella hypothesis that incorporates and constrains evolutionary explanations without resolving them.

In the last two decades, the two paradigms have begun to overlap. The discovery of shell middens contemporary with early *Homo* in Kenya has caused some anthropologists to consider more carefully the nutritional significance of aquatic resources in the Rift Valley. Shellfish, seals, and other coastal food sources also figure significantly in the early appearance of modern human behaviors (Case Study 21), and there is genetic evidence that some anatomically modern humans leaving Africa probably followed shoreline of the Indian Ocean on their way to Southeast Asia, while others may have taken a coastal route from Siberia into the Americas. Such models offer some confirmation that humans had an important relationship with the sea in the past, but they do not incorporate the assumptions about anatomical adaptations that lie at the core of the Waterside Hypotheses.

The multiplicity of versions and subhypotheses of the waterside argument makes critical evaluation of the paradigm challenging. Debating a given adaptationist argument will not be able to test or disprove the paradigm. Even determining whether a given species or genus of hominin was adapted to the water depends on the preconceptions that a person brings to the argument. It is clear that in order to communicate with paleoanthropologists, the Aquatic Ape community must engage with the fossil and archaeological record as relevant evidence, the extensive documentation of hunting and butchering terrestrial animals, and the diversity of habitats occupied.

Questions for Discussion

Q1: What is an umbrella hypothesis? What is the difference between an umbrella hypothesis and a scientific theory, such as the Theory of Evolution?

Q2: Is it necessary to be parsimonious in our understanding of the world? What are the implications of ignoring parsimony?

Q3: For most anthropologists, the Aquatic Ape Hypothesis is beneath their consideration. When should we pay attention to ideas that appear bizarre and when should we dismiss them as a waste of time?

Q4: Morgan welcomed the mantle of outside challenger of the orthodoxy and compared herself to Alfred Wegener. Why do maverick ideas have appeal for nonscientists?

Q5: What are some other examples of umbrella hypotheses in science and other disciplines?

Additional Reading

Kuliukas AV, Morgan E (2011) Aquatic scenarios in the thinking on human evolution: what they and how do they compare? In: Vaneechoutte M et al (eds) Was man more aquatic in the past? Fifty years after Alister Hardy: waterside hypotheses of human evolution. Bentham eBooks, Oak Park, pp 106–119

Langdon JH (1997) Monolithic hypotheses and parsimony in human evolution: a critique of the aquatic ape hypothesis. J Hum Evol 33:479–494

Case Study 26. What Science Is: A Cultural and Legal Challenge

Abstract The Introduction presented a standard interpretation of what science is. How important is it for us to follow that definition? Can society arbitrarily change it if it so desires? In 2004 a group of school board members aligned themselves with people who were attempting to do just that—change the definition of science to make it more consistent with their religious beliefs. The legal protest from parents went to a federal court where Judge Jones evaluated and rejected the claim that Intelligent Design qualified as science.

Intelligent Design

In 2004, the definitions of science were put into question before Federal Judge John E. Jones III, who attempted to resolve a case that pitted science versus religion. The case arose when a school board in Dover, PA, attempted to require ninth-grade biology teachers to read the following statement to their students:

> The Pennsylvania Academic Standards require students to learn about Darwin's Theory of Evolution and eventually to take a standardized test of which evolution is a part.
>
> Because Darwin's Theory is a theory, it continues to be tested as new evidence is discovered. The Theory is not a fact. Gaps in the Theory exist for which there is not evidence. A theory is defined as a well-tested explanation that unifies a broad range of observations.
>
> Intelligent Design is an explanation of the origins of life that differs from Darwin's view. The reference book "*Of Pandas and People*" is available in the library along with other resources for students who might be interested in gaining an understanding of what Intelligent Design actually involves.
>
> With respect to any theory students are required to keep an open mind. The school leaves the discussion of the Origins of Life to individual students and their families. As a Standards-driven district, class instruction focuses upon preparing students to achieve proficiency on Standards-based assessments.

Eleven parents, represented by the American Civil Liberties Union and other groups, filed a lawsuit to overturn this requirement on the grounds that it introduced specific religious concepts into the public schools. Judge Jones ruled in favor of the

J.H. Langdon, *The Science of Human Evolution*,
DOI 10.1007/978-3-319-41585-7_26

plaintiffs, writing a 139-page decision that touched on many issues, both scientific and religious.

It is appropriate briefly to examine the much-maligned word "theory." It has many different and valid definitions. One is the formal definition in science—a broadly explanatory hypothesis that has been repeatedly tested and supported so as to gain a reasonably wide acceptance in the scientific community. A second, more vernacular definition is nearly the opposite: a theory is an untested surmise. Those definitions have been deliberately confused at times to undermine scientific argument or elevate fringe ideas. Even in the statement earlier, where "theory" is used properly in its scientific sense, the distinction between theory and "fact" is being used to undermine a specific theory. Since all scientific theories are held tentatively and open to the possibility of new observations that require refinement, "fact" can only apply to observations themselves. Gravity, atomic structure, and the role of germs in causing disease are also theories, but they are now accepted unhesitatingly by scientists even though there are many unanswered details about them.

At the risk of oversimplification, the major arguments of Intelligent Design (ID) may be summarized as follows: Some aspects of life are "irreducibly complex." That is, at the biochemical level they involve so many specific components, that the absence of any one part renders the rest of the system without function. Therefore, they could not have come about by a gradual or step-by-step approach. This argument was put forth by Michael Behe, most notably in his book *Darwin's Black Box*. Researchers are able to identify such systems and recognize that they were designed because they contain a "specified complexity" that is distinguishable from randomness. Another terminology is that they contain "information" rather than "noise." The only explanation for their existence is that such systems were designed by an intelligent agency and created fully formed.

The Intelligent Design (ID) model has been offered as an alternative to evolution. Both approaches purport to explain why organisms are well adapted to their environments and to carry out the functions necessary for life. Darwin proposes natural selection as a naturalistic process that can create order and complexity. ID relies on supernatural agency. It is not the intent of this chapter to critique ID in the validity of its arguments, but to use it to explore the definition of science. To offer ID as a valid scientific hypothesis, Behe must challenge the existing definition of science to permit the inclusion of supernatural explanations.

The trial in Dover addressed many issues. Judge Jones' opinions determined that the actions of the School Board had a religious purpose and were therefore unconstitutional. He examined the scientific argument for and against the concepts of ID and determined it was not supported by science. He also considered ID in relation to the definition of science, because that had the most direct bearing on whether it was appropriate to teach it in the science classrooms. The following pages are drawn from that part of his decision that addressed the nature of science. In this text, the "plaintiff" refers to the parents and witnesses arguing for a traditional understanding of science. The "defendants" are the proponents of ID. The judge also refers to a previous 1982 court decision *Maclean vs. Arkansas Board of Education* in which a federal judge ruled that "creation science" was a religious belief and it was unconstitutional for public schools to teach it as science.

The following excerpt comes from the judge's decision in Kitzmiller v. Dover Area School District/4, page 64–89. Most internal references to trial testimony have been deleted.

After a searching review of the record and applicable caselaw, we find that while ID arguments may be true, a proposition on which the Court takes no position, ID is not science. We find that ID fails on three different levels, any one of which is sufficient to preclude a determination that ID is science. They are: (1) ID violates the centuries-old ground rules of science by invoking and permitting supernatural causation; (2) the argument of irreducible complexity, central to ID, employs the same flawed and illogical contrived dualism that doomed creation science in the 1980s; and (3) ID's negative attacks on evolution have been refuted by the scientific community. As we will discuss in more detail below, it is additionally important to note that ID has failed to gain acceptance in the scientific community, it has not generated peer-reviewed publications, nor has it been the subject of testing and research. Expert testimony reveals that since the scientific revolution of the 16th and 17th centuries, science has been limited to the search for natural causes to explain natural phenomena. This revolution entailed the rejection of the appeal to authority, and by extension, revelation, in favor of empirical evidence. Since that time period, science has been a discipline in which testability, rather than any ecclesiastical authority or philosophical coherence, has been the measure of a scientific idea's worth. In deliberately omitting theological or "ultimate" explanations for the existence or characteristics of the natural world, science does not consider issues of "meaning" and "purpose" in the world. While supernatural explanations may be important and have merit, they are not part of science. This self-imposed convention of science, which limits inquiry to testable, natural explanations about the natural world, is referred to by philosophers as "methodological naturalism" and is sometimes known as the scientific method. Methodological naturalism is a "ground rule" of science today, which requires scientists to seek explanations in the world around us based upon what we can observe, test, replicate, and verify.

The judge looked to outside experts to define science and the scientific method.

As the National Academy of Sciences (hereinafter "NAS") was recognized by experts for both parties as the "most prestigious" scientific association in this country, we will accordingly cite to its opinion where appropriate. NAS is in agreement that science is limited to empirical, observable and ultimately testable data: "Science is a particular way of knowing about the world. In science, explanations are restricted to those that can be inferred from the confirmable data—the results obtained through observations and experiments that can be substantiated by other scientists. Anything that can be observed or measured is amenable to scientific investigation. Explanations that cannot be based upon empirical evidence are not part of science."

This rigorous attachment to "natural" explanations is an essential attribute to science by definition and by convention. We are in agreement with Plaintiffs' lead expert Dr. Miller, that from a practical perspective, attributing unsolved problems about nature to causes and forces that lie outside the natural world is a "science stopper." As Dr. Miller explained, once you attribute a cause to an untestable supernatural force, a proposition that cannot be disproven, there is no reason to continue seeking natural explanations as we have our answer....

In contrast, the opinion noted that ID fails this definition by turning to supernatural explanations. In doing so, its supporters knowingly place themselves outside of science.

It is notable that defense experts' own mission, which mirrors that of the IDM itself, is to change the ground rules of science to allow supernatural causation of the natural world, which the Supreme Court in *Edwards* [*v. Aguillard*] and the court in *McLean* [*v. Arkansas*] correctly recognized as an inherently religious concept. First, defense expert Professor Fuller agreed that ID aspires to "change the ground rules" of science and lead defense expert Professor Behe admitted that his broadened definition of science, which encompasses ID, would also embrace astrology. Moreover, defense expert Professor Minnich acknowledged that for ID to be considered science, the ground rules of science have to be broadened to allow consideration of supernatural forces.

... Notably, every major scientific association that has taken a position on the issue of whether ID is science has concluded that ID is not, and cannot be considered as such. Initially, we note that NAS, the "most prestigious" scientific association in this country, views ID as follows: Creationism, intelligent design, and other claims of supernatural intervention in the origin of life or of species are not science because they are not testable by the methods of science. These claims subordinate observed data to statements based on authority, revelation, or religious belief. Documentation offered in support of these claims is typically limited to the special publications of their advocates. These publications do not offer hypotheses subject to change in light of new data, new interpretations, or demonstration of error. This contrasts with science, where any hypothesis or theory always remains subject to the possibility of rejection or modification in the light of new knowledge.

The court decision then analyzed Intelligent Design and its claims of scientific validity.

ID is at bottom premised upon a false dichotomy, namely, that to the extent evolutionary theory is discredited, ID is confirmed. This argument is not brought to this Court anew, and in fact, the same argument, termed "contrived dualism" in *McLean*, was employed by creationists in the 1980s to support "creation science." The court in *McLean* noted the "fallacious pedagogy of the two model approach" and that "[i]n efforts to establish 'evidence' in support of creation science, the defendants relied upon the same false premise as the two model approach ... all evidence which criticized evolutionary theory was proof in support of creation science." We do not find this false dichotomy any more availing to justify ID today than it was to justify creation science two decades ago.

ID proponents primarily argue for design through negative arguments against evolution, as illustrated by Professor Behe's argument that "irreducibly complex" systems cannot be produced through Darwinian, or any natural, mechanisms. However, we believe that arguments against evolution are not arguments for design. Expert testimony revealed that just because scientists cannot explain today how biological systems evolved does not mean that they cannot, and will not, be able to explain them tomorrow ... It also bears mentioning that as Dr. Miller stated, just because scientists cannot explain every evolutionary detail does not undermine its validity as a scientific theory as no theory in science is fully understood.

As referenced, the concept of irreducible complexity is ID's alleged scientific centerpiece. Irreducible complexity is a negative argument against evolution, not proof of design, a point conceded by defense expert Professor Minnich (Minnich: irreducible complexity "is not a test of intelligent design; it's a test of evolution")....

The judge summarized argument that evolutionary theory does offer explanations for the emergence of systems with complex parts. Behe attempted to dismiss such explanations by deliberately ignoring such models and the evidence supporting them.

As irreducible complexity is only a negative argument against evolution, it is refutable and accordingly testable, unlike ID, by showing that there are intermediate structures with selectable functions that could have evolved into the allegedly irreducibly complex systems. Importantly, however, the fact that the negative argument of irreducible complexity is testable does not make testable the argument for ID. Professor Behe has applied the concept of irreducible complexity to only a few select systems: (1) the bacterial flagellum; (2) the blood-clotting cascade; and (3) the immune system. Contrary to Professor Behe's assertions with respect to these few biochemical systems among the myriad existing in nature, however, Dr. Miller presented evidence, based upon peer-reviewed studies, that they are not in fact irreducibly complex....

We will now consider the purportedly "positive argument" for design encompassed in the phrase used numerous times by Professors Behe and Minnich throughout their expert testimony, which is the "purposeful arrangement of parts." Professor Behe summarized the argument as follows: We infer design when we see parts that appear to be arranged for a purpose. The strength of the inference is quantitative; the more parts that are arranged, the more intricately they interact, the stronger is our confidence in design. The appearance of design in aspects of biology is overwhelming. Since nothing other than an intelligent cause has been demonstrated to be able to yield such a strong appearance of design, Darwinian claims notwithstanding, the conclusion that the design seen in life is real design is rationally justified....

Testimony established that Behe's argument is "not scientific" and cannot be falsified. Furthermore, the basis of his reasoning is a false analogy of biological systems to human artifacts. But artifacts do not reproduce themselves and are not subject to natural selection; therefore, comparisons do not inform us about evolution.

It is readily apparent to the Court that the only attribute of design that biological systems appear to share with human artifacts is their complex appearance, i.e. if it looks complex or designed, it must have been designed. This inference to design based upon the appearance of a "purposeful arrangement of parts" is a completely subjective proposition, determined in the eye of each beholder and his/her viewpoint concerning the complexity of a system. Although both Professors Behe and Minnich assert that there is a quantitative aspect to the inference, on cross-examination they admitted that there are no quantitative criteria for determining the degree of complexity or number of parts that bespeak design, rather than a natural process. As Plaintiffs aptly submit to the Court, throughout the entire trial only one piece of evidence generated by Defendants addressed the strength of the ID inference: the argument is less plausible to those for whom God's existence is in question, and is much less plausible for those who deny God's existence.

Accordingly, the purported positive argument for ID does not satisfy the ground rules of science which require testable hypotheses based upon natural explanations. ID is reliant upon forces acting outside of the natural world, forces that we cannot see, replicate, control or test, which have produced changes in this world. While we take no position on whether such forces exist, they are simply not testable by scientific means and therefore cannot qualify as part of the scientific process or as a scientific theory.

The judge next turned to the question of whether the validity of evolution had been undermined by the claims of ID that there are serious limits in the ability of evolution to explain life. He concluded that their argument misrepresented science. They ignored supporting evidence and dismissed the overwhelming support that evolutionary theory receives from the scientific community. The Panda's textbook written to present ID to schoolchildren has numerous errors and deliberately distorts fundamental concepts such as phylogeny, homology, and exaptation. Finally, the ID community has failed to participate in the recognized community of science by offering its work to peer review and criticism.

A final indicator of how ID has failed to demonstrate scientific warrant is the complete absence of peer-reviewed publications supporting the theory. Expert testimony revealed that the peer review process is "exquisitely important" in the scientific process. It is a way for scientists to write up their empirical research and to share the work with fellow experts in the field, opening up the hypotheses to study, testing, and criticism. In fact, defense expert Professor Behe recognizes the importance of the peer review process and has written that science must "publish or perish." Peer review helps to ensure that research papers are scientifically accurately, meet the standards of the scientific method, and are relevant to other scientists in the field. Moreover, peer review involves scientists submitting a manuscript to a scientific journal in the field, journal editors soliciting critical reviews from other experts in the field and deciding whether the scientist has followed proper research procedures, employed up-to-date methods, considered and cited relevant literature and generally, whether the researcher has employed sound science. The evidence presented in this case demonstrates that ID is not supported by any peer-reviewed research, data or publications. Both Drs. Padian and Forrest testified that recent literature reviews of scientific and medical-electronic databases disclosed no studies supporting a biological concept of ID. On cross-examination, Professor Behe admitted that: "There are no peer reviewed articles by anyone advocating for intelligent design supported by pertinent experiments or calculations which provide detailed rigorous accounts of how intelligent design of any biological system occurred." Additionally, Professor Behe conceded that there are no peer-reviewed papers supporting his claims that complex molecular systems, like the bacterial flagellum, the blood-clotting cascade, and the immune system, were intelligently designed. In that regard, there are no peer-reviewed articles supporting Professor Behe's argument that certain complex molecular structures are "irreducibly complex." In addition to failing to produce papers in peer-reviewed journals, ID also features no scientific research or testing. After this searching and careful review of ID as espoused by its proponents, as elaborated upon in submissions to the Court, and as scrutinized over a 6 week trial, we find that ID is not science and cannot be adjudged a valid, accepted scientific theory as it has failed to publish in peer-reviewed journals, engage in research and testing, and gain acceptance in the scientific community. ID, as noted, is grounded in theology, not science. Accepting for the sake of argument its proponents', as well as Defendants' argument that to introduce ID to students will encourage critical thinking, it still has utterly no place in a science curriculum. Moreover, ID's backers have sought to avoid the scientific scrutiny which we have now determined that it cannot withstand by advocating that the controversy, but not ID itself, should be taught in science class. This tactic is at best disingenuous, and at worst a canard. The goal of the IDM is not to encourage critical thought, but to foment a revolution which would supplant evolutionary theory with ID.

To conclude and reiterate, we express no opinion on the ultimate veracity of ID as a supernatural explanation. However, we commend to the attention of those who are inclined to superficially consider ID to be a true "scientific" alternative to evolution without a true understanding of the concept the foregoing detailed analysis. It is our view that a reasonable, objective observer would, after reviewing both the voluminous record in this case, and our narrative, reach the inescapable conclusion that ID is an interesting theological argument, but that it is not science.

The Importance of Science

Judge Jones' decision carefully considered the definition of science. It is a mode of investigation that is based on observation. It considers only natural explanations and not the supernatural. It is constructive, seeking better ideas, and not merely criticizing. It operates within a community that encourages review, critique, and debate and ultimately determines which ideas should be supported. Natural science has been described as a way of knowing, but there are other ways of knowing. Why should science be favored in a classroom?

Science is the best means of understanding the world in principles that predict the outcomes of our actions. This is important if we want to make intentional changes in the environment, or if we want to preserve it, or if we only want to understand out own impact on it. It is essential for the development of technologies of all kinds and for our ability to influence our future as a society or as a species.

Recent decades have witnessed many attempts to confuse, hide, or deny scientific findings for monetary or religious reasons. Tobacco companies fought to prevent the public from understanding the danger cigarettes posed to their health. Energy industries have lobbied to deny the reality and the impact of climate change.

Trial attorneys spread false rumors about connections between vaccinations and autism. Religious fundamentalists continue to battle the teaching of evolution and other scientific concepts they believe threatens their faith. These fabrications deliberately sow confusion and distrust that have caused and will cause great harm and death to many people. It is of vital importance that we teach each generation how to comprehend and discern good science, and we can do that by keeping bad science out of the science classroom. That is the real importance of the trial in Dover.

Questions for Discussion

Q1:Why doesn't science study supernatural phenomena? Can supernatural explanations allow us to predict the outcomes of experiments or of the course of nature?

Q2:Judge Jones decision strongly rejects the scientific validity of ID arguments. However, in the first sentence of this passage you read, Jones states "ID may be true." In the last paragraph, he states "we express no opinion on the ultimate veracity of ID as a supernatural explanation." This is neither sarcasm nor hypocrisy. How can ID be true if it has no scientific validity?

Q3:Other academic disciplines use different rules than science. How are sociology, literature, economics, religion, and history different from science? That is, why are they not considered natural sciences?

Q4:It has been argued that the Bible constitutes evidence against evolution. Can texts represent scientific evidence?

Q5:Among the criteria Judge Jones cited to recognize legitimate science were endorsement by professional scientific bodies and publication validated through peer review. Are these reasonable expectations? Do they tend to privilege orthodox views and create an unwanted barrier to new ideas?

Q6:What are the dangers of teaching ID as an alternative scientific model to evolution?

Additional Reading

The entire Dover decision may be accessed at<en.wikisource.org/wiki/Kitzmiller_v._Dover_Area_School_District_et_al.>

Another summary of Intelligent Design, with a bibliography may be accessed ad www.designinference.com/documents/2003.08.Encyc_of_Relig.htm

Index

J.H. Langdon, *The Science of Human Evolution*,
DOI 10.1007/978-3-319-41585-7

Printed in the United States
By Bookmasters